Fundamentals of Applied Statistics and Surveys

Fundamentals of Applied Statistics and Surveys

David B. Orr

Advanced Educational Concepts

JOHNNY BLAIR

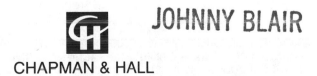

CHAPMAN & HALL

New York • Albany • Bonn • Boston • Cincinnati • Detroit • London • Madrid • Melbourne
Mexico City • Pacific Grove • Paris • San Francisco • Singapore • Tokyo • Toronto • Washington

For more information, contact:

Chapman & Hall
115 Fifth Avenue
New York, NY 10003

Chapman & Hall
2-6 Boundary Row
London SE1 8HN

International Thomson Publishing
Berkshire House 168-173
High Holborn
London WC1V 7AA
England

International Thomson Editores
Campos Eliseos 385, Piso 7
Col. Polanco
11560 Mexico D.F. Mexico

Thomas Nelson Australia
102 Dodds Street
South Merlbourne, 3205
Victoria, Australia

International Thomson Publishing Gmbh
Königwinterer Strasse 418
53228 Bonn
Germany

Nelson Canada
1120 Birchmount Road
Scarborough, Ontario
Canada, M1K 5G4

International Thomson Publishing Asia
221 Henderson Road
#05-10 Henderson Building
Singapore 0315

International Thomson Publishing-Japan
Hirakawacho-cho Kyowa Building, 3F
1-2-1 Hirakawacho-cho
Chiyoda-ku, 102 Tokyo
Japan

1 2 3 4 5 6 7 8 9 10 XXX 01 00 99 97 96 95

Library of Congress Cataloging-in-Publication Data

Orr, David B.
    Fundamentals of applied statistics and surveys : basic principles for students, teachers, managers, researchers, and other data users / by David B. Orr.
        p.   cm.
    Includes bibliographical references and index.
    ISBN 0-412-98821-6
    1. Statistical. 2. Surveys. I. Title.
    QA276.12.O77   1994
    001.4'22—dc20   94-11048
                                                                 CIP

Please send your order for this or any Chapman & Hall book to **Chapman & Hall, 29 West 35th Street, New York, NY 10001, Attn: Customer Service Department.** You may also call our Order Department at 1-212-244-3336 or fax your purchase order to 1-800-248-4724.

For a complete listing of Chapman & Hall's titles, send your requests to **Chapman & Hall, Dept. BC, 115 Fifth Avenue, New York, NY 10003.**

Contents*

Nota Bene: Detailed Tables of Contents are to be found at the beginning of each chapter. These are intended to help the reader select from among the topics those of particular interest. In addition, selected references are listed at the end of each chapter.

Preface

Purpose of This Book

This book has a broader purpose than as a teaching text intended solely to train statisticians. Its purpose is explanation and elucidation of statistical ideas. It is offered in the belief that we all live in a world in which such ideas abound, and that successful living frequently demands understanding them. Hence it is heavily focussed on conceptual content, and only incidentally on computation of the various statistical indices.

Further, many of the more complex statistical procedures of multivariate analysis have been omitted in favor of a focus on the more easily understandable, and more generally useful, techniques most likely to be encountered by those in careers other than statistics per se. Thus, its primary attraction will not be for those who intend to become professional statisticians.

However, many students are required to take statistics as a "tool" subject, intended to help them in the many career areas in which data and numerical information are important to their success. For these students, whose primary interest is the understanding of statistical ideas and processes from the standpoint of a potential *consumer* of such material, this book should prove valuable.

Researchers, administrators, and managers have similar needs and interests. For them, it is frequently not required that they design and conduct statistical studies, but rather that they buy the statistical services and products of others intelligently; or that they must understand reports and data of a statistical nature to keep up with their fields and make informed business decisions. For them, too, this book should be of considerable value.

Even members of the general public might find this book valuable as a reference. Parents, teachers, and school administrators often encounter statistical reports on student performance and other such matters. Also, surely, we are all exposed to polls, surveys, and the like daily.

Thus, the focus of this book is on users of statistical information rather than producers of such information. In cases where design and computation are required, it is advisable to secure the services of an experienced, professional statistician and I would not advise you otherwise.

To reiterate, then, this book is not intended to make a practicing professional statistician out of the reader, but rather to give non-career-statisticians a better understanding of the concepts essential to the understanding of, and intelligent application of, common statistical procedures and ideas. Perhaps a secondary benefit will be an enlarged appreciation of the extent to which "statisticalness" is an integral part of our daily lives.

Why I Decided to Write This Book

Perhaps it would be well to offer some background about how I came to undertake this effort. (If you are not interested, please skip to the next section.) Sometime about the second grade I became aware that numbers were different from other school subjects. Certainly the elementary level skills that I learned in those days were more immediately useful to me than some of the other school subjects. Being able to count things, to buy things and make change, and to understand the numbers on the speedometer of Dad's car were everyday triumphs.

However, I also became aware that arithmetic and numbers had other properties of difference. For example, elementary level teachers sometimes looked as though they had swallowed a lemon when it came time for the daily lesson in arithmetic. As I was fortunate enough not to have much trouble with my studies, I couldn't understand this.

Only much later, after I became a student of educational psychology, did I retrospectively begin to understand some of the factors at work. Clearly some of these teachers were themselves uncomfortable with the subject matter of numbers, arithmetic, and mathematics—and their discomfort was projected to their unknowing, but sensitive, students who in turn experienced some level of discomfort.

During my 20+ years of teaching statistics-related material, mostly to adults, I found that first conferences with students very frequently began with "I never was very good at math...," followed by a litany of troubles and self-doubts concerning things numerical. Unfortunately, whatever their success in my statistics courses, this negative image of their own lack of proficiency in areas involving math tended to persist.

Why should the study of numerical processes produce such anxiety? I believe that one of the critical differences between math and most other subjects, certainly those at the elementary school level, is that math is the only subject where you are usually clearly either right or wrong! And furthermore, your classmates usually know whether you are right or wrong! This goes for the teacher's performances as well as the students'! Although the teacher is an adult, being wrong in front of the class is just as traumatic, if not more so!

Consider, if an eighth grader studying American History is asked "What were the causes of the Civil War?," there are numerous answers that are, at least to some degree, correct. Also, there is an almost infinite variation in details and relationships impacting the degree of correctness that can be used to justify one's proffered answer.

One might cite "slavery," for instance; or the impact of slavery on the economic competition between the North and the South; or the underlying differences between the agricultural economy of the South and the industrial economy of the North which made slavery a key factor. Or, one might talk about an entirely different class of causes, such as the military confrontation at Fort Sumter.

In any case, although various such answers may be differentially "good," none of them is "wrong," and many others are permissible. Contrast this with, "What is the remainder when 4640 is divided by 13?" You either come up with 12, or you don't! Or, "Simplify the following:

$$\frac{5/8 + 3/4}{1/2} = ?"$$

The answer is 2 3/4, and nothing else.

Thus, it is clear that math studies can be risky. Not only is there little room for waffling, but much of the time one has to produce the answer in front of one's watching, waiting peers. For many children, and adults as well, the benefits of mathematical proficiency fall far short of the risks of exposing oneself to one's peers as incompetent, thus forcing a diminution in the esteem in which one may hold oneself.

Unfortunately, one's peers all too frequently find one's discomfiture amusing as well. It doesn't take too many experiences of this type (public failures) to produce a lasting effect, and of course no one is smart enough or lucky enough to avoid all mathematical mistakes. The natural result of all this is to generate a hesitancy toward working with numbers. Often, the education system apparently either takes little note of this situation, or lacks the insight and techniques to remedy it.

It seems likely that many of those who might find this book of some value may well harbor some of these discomforts. Also, of course, although statistics is not mathematics, like many subjects it depends to some degree on mathematical ideas and processes. Since we can't avoid this entirely, it seems best to deal with it. But the question is how.

Another well-tested psychological principle is that fears are reduced by familiarity and understanding. Because we can expect that some readers will have some fears related to mathematics that will carry over to their attempts to understand and use statistics, it is the strategy of this book to foster familiarity and understanding. To do this, I will lay a foundation for statistics that is rooted as much as possible in familiar ideas and experiences as a way to reach the basic concepts and ideas essential to the subject.

At the same time, it seems to be desirable to avoid the elements that historically stimulate the greatest fears in students. Thus, we will avoid excessive preoccupation with esoteric mathematics, formulae, and derivations. Again, wherever possible, the book attempts to call upon familiar concepts and ideas bearing on numbers and their applications to modern life.

Thus, this book is an attempt to talk about statistical/numerical ideas, using a discursive rather than a computational approach. It is conceived as an exploration of ideas—not as the traditional mathematical or statistical "cookbook," which provides computational formulae with a minimum of explanation. Indeed, one of the most memorable aspects of my own studies in these areas is the frequency with which textbook authors would offer a cryptic sentence or two and then say, "It is clear that . . ." or, "It obviously follows that . . . ," when I didn't think it was clear or obvious at all!

Another fact that has tugged at me for some years is the fact that I was some years past Introductory Statistics before some of the interpretational understandings of certain basic statistics became clear to me through my own efforts. Why wasn't more focus placed on such ideas and interpretations in the statistical texts we used? I don't know.

Again, as is true in many math books, there were always heavy emphases on mathematical derivations that were difficult and of dubious value in understanding either the ideas or the applications involved, and may indeed have usurped the time otherwise available for conceptual and exemplary elucidation.

A combination of these and other frustrations has led to this book. It is certainly not a traditional textbook, although it might be used an adjunct to some more traditionally oriented statistics text. However, it might also be applicable and useful outside of the context of statistical coursework entirely. It is my hope that it will prove to be broadly applicable, and it is this hope that has led to my decision to focus it as I have, largely on ideas rather than on computation.

Contents and Organization

As you may well imagine, it was no mean task to select the content to be covered. As one wag has put it, everything seems related to everything else —so what do you leave out??

I have chosen to assume very little, believing that there are plenty of intelligent people in this world who are not familiar with arithmetic/mathematical/statistical technique, yet have periodic needs to deal with such things. Thus, I have gone back to the most basic aspects of statistics and have tried to incorporate in the text the fundamental ideas upon which statistics is grounded. As a result, some of the early material in the book deals with decidedly nonstatistical topics. These are, however, in my opinion, fundamental to the understanding of the statistical concepts that form the major emphasis for the book.

Some of the topics that are touched on, at least briefly, include the nature of numbers, classes of numbers, applications of numbers, a few topics in algebra, the basics of expressions and their combinations, fractions, square roots, summation procedures, significant digits and rounding, measurement, scientific method and elementary experimental design. Some of these topics are needed to understand the mechanics of statistics, and others are vital to statistical applications—using statistics in everyday life, and understanding the results.

On the other hand, I have pursued these nonstatistical topics only so far as I feel is necessary for our purposes here—a decidedly limited treatment in most cases. Indeed, I have pursued most of the topics covered in the book in general only so far as I think necessary to achieve an understanding of the ideas and concepts involved.

In some cases this has resulted in a much more limited treatment (but perhaps more understandable) than is found in more traditional statistics books. [This approach, I fear, will probably also seem terribly elementary to "real" (professional) statisticians. But, never mind—they already know all of this stuff, and they can't imagine anyone having trouble understanding any of it.]

This book is organized into nine substantive chapters preceded by this Preface, and followed by a tenth chapter that acts as a brief General Summary. The topical content of these chapters is shown in some detail in the Table of Contents that precedes each chapter. The chapters follow a generally logical approach to the subject and, other things being equal, should be taken in turn, although many of the ideas presented can stand alone. The chapters are further grouped into several broad parts.

Part I

Part I, comprised of Chapters 1 and 2, covers the background and contextual material essential for understanding statistics. Statistics after all works with numbers. Therefore, it is important to know something about numbers, what they are, where they fit in, and how they are manipulated.

Chapter 2 introduces some basic concepts in statistics and provides a context for the more traditionally oriented statistical topics of later chapters. A major concern is to provide some insight into the sources of the numbers upon which one may perform statistical operations. Data, of course, are the source material for statistical applications; thus they must be assembled and manipulated in known and legitimate ways in order that the statistical application be accurate and acceptable. After all, the results of statistical procedures can be no better than the numbers upon which they were based.

Part II

Part II is comprised of Chapters 3, 4, and 5, which deal with the most fundamental of statistical treatments: distributions, averages, and variability, respectively. These topics are of most general interest and easiest for the novice to understand.

Part III

Part III deals with somewhat more complex statistical material, specifically relationships and inference. Chapters 6 and 7 are devoted to the understanding of statistical relationships and predictions based upon established relationships. Chapter 8 looks at the issues of statistical inference and its logic, hypothesis testing, and confidence.

Part IV

The final substantive portion of the book, Chapter 9, delves into one of the large-scale applications of statistics, the survey process. Surveys are in frequent use today to provide opinion data, shed light on the progress of our politics and politicians, document the progress of education, and the like.

Finally, a very brief summary is contained in a concluding Chapter 10.

In Summary

All this is not to say that everybody desperately needs this book. However, it is becoming increasingly clear that our world is becoming more and more complex. That complexity has a tendency to be reflected in statistical approaches to things that comprise our everyday experiences. It is my hope that the reader will find at least some few of the understandings that are needed to cope with such a world within these pages.

Acknowledgments

I wish to express my appreciation for the comments, criticisms, and suggestions made by several of my colleagues who reviewed earlier drafts of this book. They include the following:

Barbara J. Barrett, M.A.
   Private Consultant, Fredericksburg, VA
Marjorie O. Chandler, Ph.D.
   National Center for Education Statistics (Ret.), Fairfax, VA
David Friedman, Ph.D.
   Research Applications, Inc., Rockville, MD
Robert L. Morgan, Ph.D.
   U.S. Dept. of Education, Washington, DC
Roy C. Nehrt, M.A.
   National Center for Education Statistics (Ret.), Springfield, VA

Particularly extensive and searching critiques were conducted by two additional colleagues, both of whom deserve especial thanks:

Mary K. Batcher, Ph.D.
   Statistics of Income Division, Internal Revenue Service, Washington, DC
John L. Hayman, Jr., Ph.D.
   Troy State University at Dothan, AL (Ret.), Birmingham, AL

Finally, I should like to acknowledge Advanced Educational Concepts, Fredericksburg, VA, a recently organized nonprofit research corporation, for their generous support.

Fundamentals of Applied Statistics and Surveys

1

Foundations for the Study of Statistics

One of the hobby horses that I ride in this field is that it isn't sufficient to know something about statistics per se. Statistics could not exist without numbers (data) upon which to operate. Therefore, it is vital to know something about the number system, what it is, where it fits, and how it works. The purpose of Chapter 1 is to provide a synopsis of what numbers are and some ways in which they are manipulated. Particular emphasis is given to certain numeric procedures of special interest to statisticians, but no effort is made to cover all aspects of arithmetic / mathematical procedure.

Knowing about numbers is not enough either. It is equally imperative to understand the context within which statistics operates. What is statistics, and what is it good for? And, where did the numbers come from? Are they good numbers? How were they produced? Do they and the subsequent statistical processing of them make any sense? Chapter 2 reviews topics pertaining to these issues, including notions of measurement and scientific methods.

Some will say that this material is outside of statistics procedures and techniques. So it is, within the narrow confines of most purely statistical courses and texts. However, if the reader should be lacking in some of these basic foundations, the understanding of statistics that is the goal of subsequent chapters is likely to be significantly impaired, or at least made much more difficult.

Many readers will already have some familiarity with the material offered in Part I and wish to move on to Part II. Even so, I urge your perusal of the chapter tables of contents and a certain amount of skipping and scanning before proceeding to the more traditionally statistical material in succeeding chapters.

I

Some Basic Notions
About Numbers

I. Overview

Grist for the statistical mills is almost always in the form of numbers.
Although statistics is not the same thing as either arithmetic or mathematics, these two disciplines provide critical foundations for statistical ideas

3

and procedures. It is not necessary to be a "whiz" in arithmetic and math to be competent in statistical ideas (or even computation). But, it *is* necessary to have a general background in the basics of these subjects, and it surely helps to cultivate a sort of "number sense" when dealing with statistical problems.

One of the dilemmas I faced in writing this book was how deeply to go into background issues. Not knowing, of course, exactly who might be reading my work, I could not know how well he or she might be grounded in the fundamentals of arithmetic/mathematics practice. I was tempted originally to start at the beginning and explain addition and subtraction, going on from there. However, with some reflection (and with some prodding from my colleagues), I finally concluded that those of you who might really profit from such material would probably become lost in the later chapters, whereas those who could profit from the later chapters would never have the patience to plow through such basics.

As usual, the solution that I have adopted is a compromise. In this chapter, I begin by placing numbers in the context of concepts and ideation. This setting may be a somewhat novel approach to you. However, it is my strong feeling that numbers rarely, if ever, concern most of us except in connection with ideas, issues, problems, and real life applications. I want to stress this connection at the outset.

This is followed by some definitions and descriptions of various kinds of numbers. Next comes a brief summary of some of the basic arithmetic operations you should be familiar with. I am particularly concerned that you know about concepts such as signed numbers and how they are combined, and about literal numbers and how they are combined, but unfortunately I can't digress to cover these topics here. I conclude with a section in which I treat several areas of arithmetic procedure that are frequently found in statistical problems: estimation, spurious accuracy and rounding, square rooting, and summation procedures.

Even though I have omitted most of the basics of mathematical and arithmetic procedure, I suspect that some of you will find much of the material in this first chapter "too tame." I hope however that each of you will review closely the Table of Contents for this chapter. Use the Table of Contents to guide you in covering this material before proceeding to Chapter 2.

II. Numbers and Concepts

Statistics (and math) deals with numbers. Thus, the place to start our discussion would seem to be...

A. What is a Number?

This question is both simple and complex. It is simple in the sense that most of us "know" what a number is from an accretion of experiences collected since childhood. It is complex in the sense that setting down an actual definition requires bringing together some ideas most people probably haven't thought much about.

When you ask people what a number is you may get answers such as "Which number do you mean?" "It's, like, you know, a 1 or a 5 or 26, you know." "When you count things up that's a number." Such answers reflect real understandings of course, and are simply impromptu responses to an issue that most people spend little time thinking about.

1. Number defined

So, what *is* a number? There are various ways of defining the term, but I chose the one offered below because, for me, it emphasizes an idea that I wish to stress. That is, numbers are intimately related to ideation and thinking—not isolated and separate from normal human activity. A number is a symbol that stands for a concept. It is usually either a written or a spoken symbol. (Although numbers may exist in the context of other sensory modalities, such as tactile form, it is not necessary for our purposes to differentiate among the various sensory forms.)

Numbers probably arose as primitive man experienced the need to keep track of, or count, certain aspects of his environment. (Gorg may have known that he had some firewood stashed away, but until he could count the number of sticks, he couldn't be sure whether or not his neighbor, Florg, was stealing some when he wasn't looking.)

2. Numbers and symbol systems

Numbers taken together comprise a symbol system useful for representing ideas and concepts and their relationships.

Because the numerical system is to a significant degree embedded in the language system, the language symbols are used when numerical concepts are included in a language context. However, there are frequently numerical contexts and relationships in which an additional set of symbols (usually arabic numerals and various signs and operations indicators) is used. For some of these there are no easy language counterparts. By this I mean that many of these numerical symbols can be translated only by using long, awkward phrases of language.

A simple example of this language difficulty is the numerical expression that results from placing an exclamation point after a number (e.g., "34!"). Any person familiar with the concept of "factorials" knows immediately what this expression means; but, to put it into words: 34 times 33 times 32 times 31 times 30 times 29, and (yawn) so on ... !

Worse, many words with common language meanings *also* have highly specialized numerical meanings, and thus the opportunity for confusion between the common usages and the specialized meanings exists. For example, we have words such as "ten" and "divide" and "total" which, though fundamentally numerical in nature, clearly have common meanings in our language. Such words rarely cause confusion because their numerical meanings closely parallel their common usages.

On the other hand, words such as factor, integration, differentiation, combination, average, and variance have very specific and specialized meanings in the numerical system that are quite distinct from their common language usages (and that are quite difficult to state in ordinary language). Such terms can cause great confusion unless it is quite clear whether the meaning of the term in the numerical as opposed to the language sense is intended.

[It might be noted in passing that psychometric studies of human abilities and aptitudes tend to confirm the duality of our symbol systems. Beginning in the 1920s with the work of the English psychologists such as Charles Spearman, running through the work of the Army Aviation Psychology Program of World War II, and right up to the present time, many studies have demonstrated this duality. (See References.)]

Such studies show that people tend to differ within themselves, and with other people, in regard to verbal (language) ability and numerical ability. This finding holds true in spite of the fact that human performance is importantly related to a basic, overall general ability factor. In other words, people tend to be better at dealing with one or the other of these two basic symbol systems. (Of course, very bright people tend to be better at dealing with both than most of us are with either!)

B. Numbers as Concepts

By now you may well be asking what all this has to do with the price of booties in Botswana. Well, we have noted that numbers are symbols, and that they represent concepts. Now, we must briefly discuss the nature of a concept (or, more loosely, an idea).

I cannot offer a full treatment of the nature of concepts and their development here, but since the purpose of this treatise is focussed on understandings, rather than merely providing the usual "cookbookish" formulas, I must summarize one crucial point:

> A concept is a collection of "critical attributes." That is, it is a collection of specific characteristics that serve to distinguish that concept from other, similar concepts.

There are several broad kinds of concepts. But, in each case, we turn to the critical attributes that define the concept for clues as to just what sets of numbers (data) are to be collected (or math used, or statistics applied) in an attempt to measure, study, and analyze that concept or idea. Thus, the practical applications of the number manipulations that comprise statistics occur when we link numerical concepts with an accurate conceptualization of the various aspects and characteristics of some problem or issue of interest.

In other words, although it is true that math and statistics tend to deal with numbers, these numbers cannot be thought of in isolation from the context from which they arose. In statistics, that context is rooted in the real world and some problem(s) upon which the statistics are intended to shed light. Thus, in statistics, numbers are not concepts that exist in a vacuum, but rather are linked as data to other concepts pertaining to some real life issue. Then, it is these data that are collected that form the grist for the statistical mill.

And, it is with reference to these critical attributes and their relationships to the original concept that we make the logical linkage from the statistical results and findings to our conclusions about the subject under consideration.

In summary then, we have seen that numbers are symbols, representing concepts, and that the statistical system is a symbol system that partially overlaps the language symbol system. The manipulation of these symbol systems is simply a shorthand, an expeditious way of manipulating concepts. Statistics then involves a linking of numerical concepts with concepts characterizing the problem or issue of interest. Thus, to be clear about the formulation of numerical/statistical problems and the interpretation of findings and results, we must be careful to relate these activities to conceptualizations and ideation that are accurately rooted in the real world.

III. Basic Information About Arithmetic and Algebra

Space (and interest, I expect) does not permit extended discussion of the various mathematical and arithmetic processes, procedures, and manipulations essential to statistics. Nonetheless, in the remainder of this chapter I will mention some of the mathematical/arithmetic notions that you will need to proceed successfully through the material in the rest of the book. Generally, there is nothing required beyond a high school level. There are, however, a few topics that are of particular interest to statisticians, and these I will discuss in some detail later in the chapter.

Statistics deals with the portion of the number system that mathematics calls "real" numbers, comprised of positive and negative integers, fractions, and irrationals. In the next section I summarize and define the principal types of numbers in our system.

As for mathematical manipulations, you should be familiar with combining fractions and simplifying numerical expressions. You must be able to do the basic arithmetic operations necessary to combine numbers algebraically, that is, taking their signs into account. You must understand what literal numbers are (discussed at length in the next section), and how they can be subjected to numerical manipulation. Be sure to use the reference list if you need help with these and related topics.

IV. An Inventory of Number Types

As I have pointed out before, the purpose of statistics is to deal with numbers (data). Therefore, in order to become knowledgeable about statistics, it is necessary to know about numbers.

Suppose we think of the set of all possible numbers that conform to the standard requirements of everyday arithmetic and mathematics. There are several ways in which this comprehensive set of numbers can be classified mathematically. We need not dwell on these here, but I will summarize them briefly.

The first division is "real" numbers vs. "imaginary" numbers. As statisticians are rarely concerned with the latter, we will confine our discussions in this book to the set of real numbers. (The interested reader is referred to any standard college algebra text for a treatment of imaginary numbers; see References.)

The real numbers can be divided into the set of positive numbers and the set of negative numbers. The positive number set can be further subdivided into three parts, the integers (whole numbers), the fractions (each comprised of the ratio of two whole numbers), and the irrational numbers (to be defined below). The negative number set can be similarly subdivided into the same three subcategories.

A. Definitions of the Various Number Sets

To avoid boring some of you, I will present brief definitions of the number types I have mentioned above in the accompanying Box, so that most of you can continue in the text below.

---

1. Positive/Negative Numbers

The positive number set is that set of numbers characterized by a positive or "plus" sign (+) preceding them. They are numbers that represent "add to," "presence of," or "gain." (It should be noted that any number that appears with no sign in front of it, such as 315, or 9, is understood to be positive, just as though it were preceded by a " + ".)

The negative number set is that set of numbers characterized by a negative or "minus" (−) sign preceding them. They are numbers that represent "take away," or "absence of," or "loss." No number is interpreted as negative unless it is preceded by a minus.

2. Integers (Whole Numbers)

A good way to think of integers is to think of them as the numbers we use in counting whole, complete, or indivisible things — 1, 2, 5, 256, 324, and so forth. For example, if there were 32 pupils in a class, or 9 marbles in a jar, 32 and 9 would be whole numbers. In most cases in statistics there will be something particular that is being examined, and we will know *what* is being counted or measured. In other words there will be a "unit of measurement" — marbles, people, yards, quarts, acres, books, apples, dollars, etc. When a whole number is coupled with a unit, it tells you how many complete units you have.

3. Fractions

When we need to talk about portions of a whole unit, we are dealing with "fractions." For example, if my mother baked a pie and cut it into 6 pieces, and my brother took one, how much pie would be left? There would be a fraction left, 5/6 in this case, since my brother took a fraction, 1/6.

Fractions are derived from whole numbers. Operationally, *a fraction is always simply the ratio of two whole numbers (one divided by the other).* In the division the whole number for the part (numerator) is written above a horizontal (or slant) line. It is divided by the whole number representing a complete unit (denominator) which is written below the line. [It is well to remember that the line (/) in any fraction is always translated: "divided by." Thus, reading from top down, 13/32 translates to 13 divided by 32, 7/5 is 7 divided by 5, and 45/67 is 45 divided by 67.]

In passing, note that fractions can be what is called "improper." This simply means that the numerator is larger than the denominator, showing more than one unit of the size indicated by the denominator. Such fractions can be expressed as "compound numbers" comprised of a number of whole units combined with a "proper" fraction representing the part of a unit left over. For example, the fraction 7/3 is improper, and translates to the compound number, 2 1/3.

Sometimes, however, it is desirable to translate the fraction into a decimal form. *Every fraction can be translated into a decimal equivalent called a decimal fraction.*

A decimal fraction is created simply by actually carrying out the division called for by the line in the fraction. For example, if we take 3/4, which means 3 divided by 4, and actually divide, we will get .75, followed by as many zeroes as you want (it "comes out even"); 15/16 becomes .9375 (it also comes out even, if you go a little farther); 5/6 is .8333 (with as many more 3s as you feel like having — it never comes out even).

4. *Rational Numbers*

All fractions are termed "rational." (This has nothing to do with whether or not they make sense; the reference is to "ratio." You will remember that fractions are defined as the *ratio* of two integers.) Note also that fractions can also be thought of as *including* the integers, because any integer can be written as itself divided by the integer 1; for example, 2/1 is a fraction that equals 2; 313/1 is a fraction that equals 313, and so on. It follows then that *integers are also included under the term, "rational."*

Further, it can be shown (but we will not; see an algebra book from the reference list) that all rational numbers have decimal equivalents that either "terminate" (come out even at some point) or "repeat" (some set of digits recurs in cycles for as long as you wish to carry out the division). In the examples above, 3/4 terminates (comes out even at .75), while 5/6 repeats (here the 3 repeats). Another example of a repeating decimal would be .272727 . . . , which arises from the fraction 3/11.

5. *Irrational Numbers*

Although it might seem that the integers and the fractions should together cover all of the possible positive (or negative) numbers, this is not so. There are also some numbers that cannot be expressed as the ratio of two whole numbers, and these are termed "irrational" (*not* formed from the *ratio* of two whole numbers). For example, it can be shown that the root of a number that is not the same power as the root is irrational. A simple example is the square root of 2. The square root of two is a *non*-repeating decimal which cannot be expressed as the ratio of any two integers; its value is 1.4142136 . . . , never repeating and never coming out even."

Even this rooting procedure does not exhaust the supply of irrational numbers. Another example of an irrational number is the constant pi, which is obtained when the circumference of any circle is divided by its diameter. Pi (3.14159...), as well as all other irrational numbers, are nonrepeating decimals. Again, reference to a standard algebra text will provide more detail about irrational numbers.

6. *The Number Continuum*

Taken together, the entire set of positive real numbers can be thought of as comprising a continuous dimension or line starting at zero and ranging through every conceivable value to the right up to positive infinity. Similarly, the negative set ranges continuously through all possible negative values leftward to negative infinity.

B. Literal (Algebraic) Numbers

There is one additional type of number that is extremely important to statistics, so I must discuss it in some detail before we can proceed. This is a type found in algebra that is called a "literal" number.

Let's work up to the idea of literal number by starting with the following situation. Suppose I get a government job and they tell me to go to Calabash County and count the sheep on the various farms there. I would get something like this:

$$
\begin{array}{ll}
\text{Jones Farm} & 10\ \text{Sheep} \\
\text{Smith Farm} & 23\ \text{Sheep} \\
\text{Brown Farm} & 14\ \text{Sheep} \\
\text{Hauck Farm} & 7\ \text{Sheep} \\
\hline
\text{Total} & 54\ \text{Sheep}
\end{array}
$$

After a while, I would probably get tired of writing out "Sheep," and my tally sheet would look like this:

$$
\begin{array}{ll}
\text{Jones} & 10s \\
\text{Smith} & 23s \\
\text{Brown} & 14s \\
\text{Hauck} & 7s \\
& 54s
\end{array}
$$

Although I dropped the sheep in favor of "s," which of course stands for sheep, I cannot drop the "s" too, since next week I know I am going to have to count cows (which, now that I am experienced, I will record as "c," for cows), and I don't want people to look at my tallies afterwards and say that I can't tell a cow from a sheep!

Returning to the office, I look over the reports that have been coming in from other agents. I see one that begins:

Jones...............3t
Smith...............5t
Brown..............2t
and so forth ....

As I don't know what the other agent was sent out to count, I don't know what the "t" stands for, but I can still do arithmetic operations on the data which have been supplied in this report. For example, I know that among them Smith, Jones, and Brown have 10t (addition), and that Smith has 2 more than Jones, and 3 more than Brown (subtraction), and that if Hauck had 6t, he would have twice as many as Jones (multiplication), etc.

Now in the above case we are simply using "t" as a sort of abbreviation to stand for tractor, but we can and do take this sort of thing one step further. We could let the letter be a general symbol for the individual values of the things being counted. Thus, we could let "t" stand for *number of tractors*. Then t would take the value of 3 for Jones, 5 for Smith, 2 for Brown, and so on. *In algebra (or statistics) we would call this "t" a "literal number."* Such literal numbers are frequently used to represent "variables" (which means that the value for which t stands differs, or varies, from case to case).

The term "literal number" means using a letter to represent some presently unknown number (literal is a fancy way of saying letter). Also, importantly, *we can continue to perform basic arithmetic functions with such numbers.*

Why bother with literal numbers? Using literal numbers enables us to talk about processes in general, about abstractions, and to state principles without having to specify particulars. This is very handy to mathematicians and statisticians in their work, and permits them to state theorems and propositions that can then be checked against evidence gathered at the particular level.

It is also the basis for the much maligned "formula." A formula is simply a generalized statement of a relationship among two or more aspects of a situation, using literal numbers to achieve a generality that would be impossible when using specific values.

Consider the relationship between the circumference and the diameter of a circle—any circle. We could say that for circle "A," its diameter is 3.2 and its circumference is 10.0. But, it is possible to convey the meaning that there is a relationship between these two that is true for *all* circles ONLY by using a formula showing these two factors as literal numbers: $C = \pi D$; that is, the circumference is always equal to the diameter times a constant known as pi (π), *whatever* the circle!

It is also handy to statisticians for the same reasons. As we get into the nature of various statistics in the next few chapters, it becomes much more convenient to state relationships in general terms, thus implying that they hold for all particular relationships, than to state them in particulars. As an example, consider the arithmetic mean or average. $Mean_x = \Sigma x / N$, regardless of what is being averaged. (Fear not, we'll explain that one fully in a later chapter!)

Although literal numbers are often chosen to remind us of the variable being studied, such as W for weight and H for height, this is not at all necessary. In fact, we often have need to work with literal numbers where we may not have the foggiest notion of what the letters stand for. This should be no barrier. Simply assume that they stand for something, each unique to the other, and proceed.

It is important to remember one thing in working with literal numbers: Whenever you have a compound literal number comprised of a letter with a numerical prefix, such as $5y$, *the fact that the number and the letter are written together in juxtaposition means multiplication*; thus, in the compound literal number, $5y$, if we later find out that $y = 7$, then $5y$ equals 5 *times* 7, or 35.

Performing arithmetic operations on literal numbers follows the same basic rules of signed numbers and algebraic procedure as are applicable to regular numbers.

To recapitulate a bit, we have reviewed literal numbers, as a foundation for our consideration of statistical topics in the next few chapters. Literal numbers are subject to basic arithmetic operations. Literal numbers are important because they enable us to make general statements about the relationships among things without getting bogged down in the narrowness of the particulars of specific situations. Such relationships are often set down under the name of formulas. Thus, formulas merely express the relationships among things generally, without being tied down to specific values or particular instances.

V. Notions and Procedures of Special Interest to Statisticians

There are several numerical procedures that I regard as having special importance for statisticians. The first is more of an attitude than a procedure, but is of great importance to anyone who wishes to work in quantitative areas. The others are more specific.

A. Estimation

I suppose that the real underlying dimension is one of being aware of what is going on in your work. This is a little harder to do in statistical applications than it is in some other fields. The major reason is that statistics typically deals with masses of data. In such situations one's intuition about what the data show tends to get overwhelmed in the welter of information. In addition, such conditions act to obscure errors and inconsistencies, as these too tend to get lost in the mass of information being processed.

The major safeguard against being wildly misled is simply to know what ought to be coming out of your analysis. Operationally, this is accomplished by making estimates, regularly and often, of what the results of the statistical procedures, whatever they may be, *should* be showing you. Whenever your estimates differ significantly from your actual results, it should alert you that something is probably wrong. Either your thinking about the subject is in error, or some kinds of mistakes in data collection and/or processing must be concealed in the results.

Either way, it is never good enough simply to say, "Well, that's just the way it came off the computer!" You must then take whatever steps are necessary to resolve the discrepancies.

B. Squares and Square Roots

The process of estimation can be illustrated by considering the topic of squares and square roots. Statistics is full of squares and square roots. It is sufficiently easy to make a mistake in working with these that I feel it is important to review briefly what a square root is, and how to estimate one. While most of us now own hand-held calculators that will give us either a square or a square root with the push of a button, it is still advisable to have some sense of what these values are and how they can be obtained if one's batteries run down!

The square of a number is defined as the product of that number times itself. It is indicated by placing a superscript 2 directly after the number. For example, 4 squared is indicated as 4^2, and is equal to 16 (4×4). Obviously squaring is a special case of multiplication.

The square root of a number (indicated by use of the symbol "$\sqrt{\ }$") is defined as a number that, when divided into the original number, yields itself as the quotient. Square rooting is thus the inverse of squaring. Each can be used to undo the effect of the other.

For each decimal position in a root, there will be a pair of decimal positions in its square. Why is this so? Well, think of all the numbers from 1 to just under 10. As we multiply them by themselves they give us a series of squares from one to just under 100. If we think of any single-digit square as being an "incomplete pair" that can be completed by writing a 0 in front of it (which doesn't change its value in any way), then we can say that all single digits when squared result in two digits.

This same argument holds for two-digit numbers, which uniformly (allowing for the same sort of incomplete pairs) result in four-digit squares. So the square of 2 is 04, and the square of 3 is 09, and the square of 17 is 0289, and so forth.

This property of numbers comes in very handy if we need to estimate the square root of a large, unfamiliar number. Suppose we wanted the square root of 1732. What do we know? Well, first we know that it is going to have *two* positions to the left of the decimal. Why? Because there are two *pairs* of digits to the left of the decimal in the original number.

Further, we can get a rough estimate of the square root simply by estimating the square root of the first pair; then we can let a zero stand for the second decimal position in order to preserve the proper decimal relationship.

Going back to 1732, the largest estimated square root that would go into the first pair would be 4 ($4^2 = 16$). Then we would use 0 for the second digit (for the second pair in 1732). So, our first estimate of the square root of 1732 would be 40. We can easily see that $40^2 = 1600$, and that 50^2 would be 2500. We now can say that the root lies between 40 and 50. (It is closer to the former, since 1732 is much closer to 1600 than to 2500, but careful, this is not a linear relationship.)

Thus, you see that it is easy to develop a crude estimate, or order of magnitude, for the required square root. It is good numerical practice to do this in all cases, even if you are computing the root with a calculator, in order to maintain the sense that numerical results are as expected. When you find a large discrepancy between a calculation and your expectation, this should be a red flag that there are probably errors somewhere.

(The estimation of a square root as described above can also be used as the basis for a calculation procedure. In brief, the estimated root is used to divide into the number and the difference between the quotient and the divisor is split and added to the smaller of the two to form a new estimate of the root. Two or three cycles of such divisions can produce a root, accurate to several places.)

The above discussion applies to *any positive number*. (The square roots of negative numbers are not generally used in statistics, and are usually found only in theoretical mathematical problems.)

C. Spurious Accuracy

Statisticians sometimes tell the following story to illustrate the topic of "spurious accuracy." A farmer was being interviewed by a big-city reporter doing a "local color" story. The conversation got around to the mountains surrounding the farm. The reporter asked how old the farmer thought these mountains were, and the farmer answered, "Three billion and four and a half years old." When queried further, it turned out that he had seen a geological report in the local paper estimating the age of the mountains at 3 billion years—and that that had been 4 1/2 years ago! Although this interpretation sounds plausible on the surface, it should make you feel a little uneasy.

Let's take a closer look. The farmer stated that the mountains were 3,000,000,004.5 years old. He got his figure by adding 3,000,000,000 and 4.5 together. What's unsettling about it is the extreme disparity in precision between the two figures that were added. It is reasonable to presume that the farmer's recollection for the publication date of the report was relatively accurate—to within a year, say.

However, geologists and geophysicists dealing with the age of mountains have no such precision available. If they got within a few million years of the true age of the mountains, they would probably be highly pleased. Therefore adding these two figures together is meaningless—the precision of the one is swallowed up in the imprecision of the other.

The general rule is that whenever numbers are to be combined in any of the arithmetic (or other) operations, they must be of generally the same level of precision. Otherwise, the result is imprecise, and worse, often appears more precise than it really is (and, so, is misleading).

The story about the age of the mountains, above, illustrates this latter point also. The given age of 3,000,000,004.5 years seems to be accurate at least to the nearest half year. So, not only is the accurate measurement of 4.5 years since the report overwhelmed by the imprecision of the geological estimate, but adding in the farmer's figure to the total has given the total the false appearance of a high degree of accuracy where the true level of accuracy is much less. This latter phenomenon is called "spurious accuracy."

We find examples of spurious accuracy constantly in everyday life. This is due to a combination of circumstances. People have an almost holy respect for numbers, and tend to take them at face value without critical assessment of their sources and meanings. Additionally, people are often unaware of the limitations in reported statistics and measurements. And, frequently they lack the training to know how to treat such problems, even if they recognize an element of spurious accuracy in a report.

For example, the Statistical Abstract of the United States, 1980, reports the total population of the United States for 1970 as 207,976,452. Ridiculous! Just during the time it took me to write the last sentence people died and other people were born. Who knows—or could ever know—exactly, to the last person, how many people there were in the United States on April 1, 1970. Even assuming that the Census had succeeded in making a complete count (which they didn't), there would have been substantial intraday variation in the true population throughout the benchmark day, April 1.

As I said, we get this problem of spurious accuracy all of the time— mostly from a failure or refusal to recognize that many of the things we like to keep track of are constantly in a state of flux. Can you tell me your own net worth, to the penny? I'd like to wager a lot that you can't, but unfortunately I wouldn't be able to prove my case, because I couldn't count it to the penny any better than you can!

Does this mean that it is unreasonable to ask questions such as What is the population? and, What is my net worth? Not at all. It means, however, that we must assess the level of precision that is available to us, and stay within it, so that we don't fool ourselves with "spurious accuracy." Population questions, for instance, might be answerable, meaningfully, at a level of "the nearest 1000." In fact, many of the tables reported in the Statistical Abstracts are reported to the nearest thousand. The population of the United States given above would be reported this way as 207,976,000. This brings us to the topic of how to deal with questions of spurious accuracy and differential precision in numbers in a consistent and systematic way.

D. Significant Digits

The concept of "significant digits" is intended to reveal the accuracy of any number by telling us which of the reported digits in the number can be assumed to be precise. Where does this information come from? It can only come from the gatherer/processor of the data, but once available it helps the user to assess the information and serves as a guide when numbers are to be combined. I'll use the next several paragraphs to explain how this useful process works. But, I warn you, one of the trickiest parts of the process is concerned with the fact that not all zeros are the same.

Let's go back to the population figure I gave you, 207,976,452. If I put this in extended form by adding some zeros, *without* changing its value, I could write it as follows:

$$\ldots 000,000,207,976,452.000,000,000 \ldots .$$

Now, we made the point that probably the best measurement that could be made would be to the nearest thousand people. Therefore, the digits in the units, tens, and hundreds places have no real meaning, and might be anything, as far as we know. They are *not significant*. On the other hand, we do need *something* there in order to preserve the size of the number and the relationship of the other digits to the decimal point. So, we substitute "placeholder zeros" (nonsignificant) for the 4, 5, and 2 in the hundredths, tens, and units positions. This gives us:

$$\ldots 000,000,207,976,000.000,000,000 \ldots$$

Because we *can* count thousands, the 6 in the thousands place *is* meaningful (significant); and of course, because all of the digits to its left represent multiples of thousands, they too are significant (including the 0 between the 2 and the 7). By implication, all "leading" zeros are also significant, because they represent *precisely none* in each of those decimal positions.

On the other hand, for practical reasons, we can't include the leading zeros in a count of the significant digits in this number, because we could add as many leading zeros as we wished without changing the value of the number. Thus, although technically significant, by convention the leading zeros are not included in the count, and we say that this number has *6 significant digits*.

There are two ways to express the fact that the population number has 6 significant digits. The first is to write the number as given, but to underline the smallest significant digit: 207,97<u>6</u>,452. Some folks find this confusing, or miss the underline altogether.

The second way is just to drop the nonsignificant digits (by "rounding," which we'll discuss in a minute). When we do this, we insert (nonsignificant) "placeholder" zeros for each of the dropped digits between the significant ones to the left of the decimal point and the decimal point. This gives: 207,976,000. Note that the placeholder zeros are *not* significant themselves, but without them we would not be able to distinguish between this number and 207,976., which of course is unacceptable.

To summarize, significant digits are the digits in the number that represent meaningful measurement (ignoring the infinite number of "leading" zeros that might be written in front of the first significant digit). If

there are any nonsignificant decimal positions *between* the smallest significant decimal position and the decimal point, these must be replaced with
nonsignificant "placeholder" zeros (in order to maintain the magnitude of
the number).

Any nonsignificant positions to the right of the decimal point may
simply be dropped. [However, when dropping digits, "rounding" (discussed below) should be employed.] Because this is true, when you do find
a "0" as the last digit in a number to the right of the decimal point, it is
assumed (by convention) that it is significant and really means exactly none
in that position. For example, in the number 96.10, we would count four
significant digits; since if the final zero were not significant, there would be
no reason for putting it there—96.1 would do just as well!

E. Rounding

Before examining some specific applications of the principles of significant
digits and related issues, we must digress to deal with the topic of
"rounding." Rounding is the process of adjusting the remaining digits
when, for whatever reason, it is desirable to drop some digits from a
number.

Suppose we do a division problem that doesn't come out even. When do
we quit adding places to the quotient? Well, you should have a reason for
the number of places you want to retain, either based on theoretical or
practical grounds. This reason could be rooted in the number of significant
digits that are justified by the measurement process; it could be the desire
for consistency with other numbers in the report, or it could be simply
convenience.

In any case, the rule for rounding is: Go to the next decimal place
beyond (to the right of) the place that you want to keep, and round back;
if you are given a longer number (need to round back more than one
place), take whatever digits are to be discarded and round back based
upon the entire string of digits at once—not one at a time.

Just how do you do this? The common sense rule is simple: if the digit
in the decimal position to be dropped is greater than 5, you add 1 to the
preceding position, and if it is less than 5, you simply drop it; 45.16
rounded to one decimal place is 45.2, and 45.14 rounded to one decimal
place would be 45.1!

The idea is that if the information in the places to be discarded leads
you to believe that the last digit to be retained is leaning a little up or
down, as the case may be, you recognize that the probabilities are with you
if you follow that leaning. In other words, the discarded digits are thought
of not as no information at all, but rather as information too imprecise to

be given full weight, but enough to give a clue as to direction, up or down, for the preceding decimal position.

All of this is very reasonable, but what happens when the digit to be dropped is exactly 5? In this case, there is no clue, or leaning, to rely on, and the probabilities appear to be equal as to up or down in the preceding position. Therefore, it is best to adopt a procedure that, in effect, randomizes the outcome, that is, takes it up half the time and down half the time. This is accomplished by rounding up when the preceding digit is odd, and down when it is even.

The logic of this procedure is simply that the oddness or evenness of the preceding digit should be totally random with respect to the digit to be dropped, and thus following this procedure should average 50% ups and 50% downs taken over a lot of roundings. So, 46.5 rounded to the nearest whole number rounds (down) to 46 even, and 47.5 rounds (up) to 48.

Now let's take a more complex rounding example. Consider the fraction, $17 \div 13 = 1.3076923\ldots$

Suppose we wish to retain three places to the right of the decimal for this result; the answer is 1.308. In dropping the 6923, we consider it to be 6923/10000th of one count in the 1/1000ths position. Because this is greater than a half count, we add 1 to the 7 in that position, making it an 8. This is consistent with the principle that more than half should result in rounding up.

But, it is important to do this as a single operation, not one position at a time. For example, if we had the number 14.3347 to be rounded to two decimal places (shorthand for saying "two places to the right of the decimal point"), and we did it one place at a time we would get 14.335 as the first step, and 14.34 as the second. However, if we do it correctly, we consider the discarded digits to be 47/100 of one count in the 1/100ths position of the number, and, being less than half, it is discarded. So, 14.33 is the correct answer, not 14.34!

F. Significant Digits and Rounding

Now let's look at some examples of problems involving significant digits and rounding. First, a guiding principle: *Whenever an arithmetic process results in combining precision with imprecision, imprecision is the result.*

To take an extreme example, suppose there were 10 men in a room and you were asked to estimate the total wealth represented in that room. Nine of the men gave you a financial statement, but the 10th refused to tell you even how much he had in his pockets at the time. You add up the worth of the 9, getting $310,535. You suspect spurious accuracy, and round this figure off to $311,000. Although this is good as far as it goes, you haven't answered the original question, and if you have to answer it at this point, you must say "Totally indeterminate."

No, you can't use the figure you have determined as an estimate—not even as a minimum estimate, because your 10th man may have a net worth of *anything at all, plus or minus,* and the total wealth represented in the room is a complete mystery!

Suppose, after much cajoling, your 10th man breaks down and gives you a figure. But, says he, "I don't keep records, and my best guess is about $300,000." Now you can answer the question. But, the correct answer is $600,000, not $611,000.

Note that as stated we must assume that the 10th man's data are accurate only to the nearest $100,000. This means that, according to rounding principles, his true net worth is somewhere between $250,000 and $350,000, or $300,000 ± (plus or minus) $50,000. Imprecision of that degree in the tens of thousands position obviously overwhelms the $11,000 in that position that we had from the previous nine persons. Thus, we can state our findings only to the nearest hundred thousand, or $600,000.

To restate, *no matter how accurate some of the data may be, when they are combined with less accurate data, the result is at the lower level of accuracy.*

Take the addition problem:

3601.	[4 significant digits (sd)—included zero is significant]
74.	[0074—2 sd—but leading 0s hold significant places]
1.20	[0001.20—3 sd—last zero significant, or would drop off]
500.	[0500—1 sd—as shown, last 2 0s are placeholders]
4176.20	[Rounds to 4200.—2 sd—tens, units, tenths, and hundredths places combine accuracy with inaccuracy, and are inaccurate]

Notice that the number of significant digits in the result cannot be stated as a simple rule such as the least number of significant digits in the problem (500 has only 1 sd, but the answer has 2!). It is necessary to look at each decimal position and determine whether or not imprecision is present.

Subtraction of course works in exactly the same way.

In multiplication, it is again necessary to examine what has happened in terms of the various decimal positions in the problem as the numbers are being multiplied. Suppose you take 50×43 (1 sd $\times$ 2 sd):

$$
\begin{array}{r}
50 \quad \text{(1 sd)} \\
\times \quad 43 \quad \text{(2 sd)} \\
\hline
150 \\
200 \\
\hline
2150
\end{array}
$$

2150   [Notice that the first addition column contains imprecision (3×0); the second contains the imprecision 4×0;
therefore, neither of these columns can result in a
significant digit in the answer; thus, the result is rounded
to 2 sd, 2200]

Division, being the inverse of multiplication, can be analyzed in the
same way. Consider the problem of $30 \div 14$ (1 sd $\div$ 2). We can rewrite
this problem as $(30)(1/14)$; if we evaluate $1/14$, we get $0.0714286\ldots$.
Because the 1 is the result of inverting 14, it is a pure number, that is, it is
exactly 1, and there is no question of precision; the 14, on the other hand,
as written, must be assumed to be two significant figures. (However, in
decimal form we must credit the zero between the decimal point and the 7
as significant when we involve it in the multiplication process.) So, $1/14$
rounds to 0.071 (3 sd), which is to be multiplied by 30, or vice versa:

.071   [3 sd—the leading zero holds a significant place]

30   [1 sd—the zero is a spacer only]

000   [resulting from the 0 in 30, all three are nonsignificant]

213   [all are significant]

2.130   [1 sd, rounds to 2.—3 columns imprecise]

To summarize, many times we run across data and reports that are
presented at a level of apparent numerical accuracy that is suspicious.
Logical analysis of the phenomenon being studied, the measurement
processes, and other factors warn that to take these data at face value is to
be misled. Thus, we turn to rounding to eliminate the spurious accuracy
implied by such results.

If circumstances are such that it can be done, we evaluate the data
derived from different sources for accuracy. Frequently we find vastly
different levels of accuracy are involved in the same problem or study. We
should always refrain from combining highly disparate levels of precision
in the same numerical processes and arithmetic operations. If such be-

comes necessary, the overall level of precision of the results and conclusions can be assessed by following the principles of significant digits in order to track the impact of varying degrees of precision of the data on the final results.

G. Summations

A final topic to be considered in this chapter of arithmetic basics is the set of procedures referred to as "summations." Summations are very important in statistics because of the fact that statistics usually deals with large quantities of data that must be aggregated or summed in various ways in order to extract information from them.

1. Some terms defined

In order to discuss this matter it is necessary to digress to establish the meanings of two terms to be used here. These are "constant" and "variable." You must remember that statistics exists to process large quantities of data. Most frequently these data are derived from measuring objects, people, or events, so that there is a class of things, the individual members of which are the object of study. The terms "constant" and "variable" refer to the measures derived from the individuals in the group of things being studied.

If those individuals all have the same value for a measurement, that measure is called a "constant." If the values change, or vary, from individual to individual, the measure is called a "variable."

For example, if the individuals are people, and if the measure is number of arms, the measure is a constant, and its value is 2 [there are of course a few exceptions in practice, but theoretically the value is 2]. On the other hand, if we were talking about a measure of income, or intelligence, or marital status, these would vary from person to person, and thus would be "variables."

Many statistical measures are based upon the collection of data from the individuals in a class under study, and the subsequent "summation" (adding up) of the data on the measure across all of the individuals from whom data were collected. It is thus useful to be able to express these summations in a convenient way. The most important symbol in this area is "Σ," which is a capital Greek "S" (for Sum), and which means to add up all of the values of the variable that follows the symbol. Literal numbers are used to establish formulas and summations are conducted as called for by the relationships for which the formulas stand.

2. Principles of summation

Summations are found in combination with variables. Thus, "Σx" means to add up all of the values of the variable measure for which "x" stands, across all of the individuals in the group measured. Understand that "x" is the variable, a literal number representing some characteristic of the individuals in the group of interest, which takes particular and specific values for each individual. We use "x" in order to be able to talk about these values in a general fashion. The "Σ" is an "operator" symbol, similar to "$+$" or "$\div$." It is not itself a variable, but simply tells what you must do (sum) with respect to the variable that follows it.

[It is understood when "Σ" is used that the values to be summed include those for all of the individuals in the group. This number of individuals is frequently designated "N" (for the *number* of individuals in the group). Where there is some question as to what is being summed up for which individuals, additional notation is added. For example, to make it explicit that you want to sum all individuals in the group, the subscript "i" (for individual) may be added to the "x," that is, x_i, and the instruction $i = 1$ written below the Σ and N above it. This signifies the summation of all individuals from 1 to N. (See below, left.) A subset is indicated by substituting some other appropriate instructions, written above and below the Σ. (See below, right, which indicates the summation only of individuals 3 through 9).]

$$\sum_{i=1}^{N} x_i \qquad \sum_{i=3}^{9} x_i$$

What about something like "Σxy"? Well, the Σ is only an operator, so that the thing to be added is something represented by xy. [Note that, because of the juxtaposition, this means x times y.] What this says is that we have *two* variables in this expression, and that for each of the individuals in the data base (group being studied) we must take the value for the x variable and multiply it by the value for the y variable; this produces a new variable, xy, which is the thing to be added up across the population of individuals.

Similarly, if we had "$\Sigma x/y$," you would first take the ratio of x and y for each individual in the group and then add these up for all of the individuals in the group.

Note that whenever a variable (which changes from individual to individual) is multiplied by (or subjected to any other basic arithmetic operation) EITHER a constant or another variable, the result is always a variable. Thus, if you have something that changes from individual to individual, operating on it (adding, subtracting, dividing, multiplying, or rooting) with something else that also changes won't eliminate change from the result! Even if you operate on it with something that doesn't change (a constant), the result is variable. [Think about 100 people with coins in their pockets; if you take half their coins away (dividing by the constant of 2), they still differ (vary) with respect to the coins they have left.]

Now consider "$\Sigma x \Sigma y$." Remember about things in juxtaposition. This says take the sum you get from adding the xs and multiply it times the sum you get from adding the ys. It might also be written $(\Sigma x)(\Sigma y)$; *it is different from and NOT equal to* Σxy! Similarly $\Sigma x/y$ is generally NOT equal to $\Sigma x/\Sigma y$.

A moment's thought will show that the "product of the sums" should not be expected to be equal to the "sum of the products." Suppose we have the following data set of three cases:

	x	y	xy
Case #1	3	2	6
Case #2	1	4	4
Case #3	5	6	30
	9	12	40

Note first that the product of the sums is 9 times 12, or 108; while the sum of the products is 40—clearly not the same. This arises because any given x contributes only x amount to the sum of the xs; but, its contribution to the sum of the xys depends upon which y it is paired with. Thus, while rearranging the xs makes no difference to the sum of the xs, it would make considerable difference to the sum of the xys. Note that the xs add to 9 regardless of what order you write them down. But, if you should interchange the 1 and the 5, the xy products become 20 and 6, and the sum of the xy becomes 32 instead of 40!

In other words, the sum of the numbers is independent of their order; thus, the product of the sums is independent of order. But, the sum of the products is dependent on the order of the individual xs and ys. Hence the two products will ordinarily not agree.

How about $\Sigma(x + y)$? This says that you get an x and a y from each of the individuals in the group, add the x and y, and then add up all of these sums across the group. However, note that the effect of $\Sigma(x + y)$ is that each individual x and each individual y gets included in the final sum. Because it doesn't make any difference in addition which way you group the items to be added, $\Sigma(x + y)$ is exactly equal to $\Sigma x + \Sigma y$! That is, you could add them alternately, x and y, or serially, all the xs followed by all the ys, and you would come out with the same sum.

A little thought shows that the same thing is true for $\Sigma(x - y)$. Again, all of the plus quantities are combined with all of the minus quantities into a total summation, and whichever way you group them, individually as $(x - y)$s and add, or all of the pluses then all of the minuses, you still get the same result. Thus, again, $\Sigma(x - y)$ is exactly equal to $\Sigma x - \Sigma y$!

To prove this to yourself, look at the table of numbers below and check the relationships of the various sums:

Case #	x	y	$x + y$	$x - y$
1	3	4	7	-1
2	4	1	5	3
3	6	5	11	1
4	3	6	9	-3
5	2	2	4	0
6	5	3	8	2
	23	21	44	2

As you can see, $\Sigma x + \Sigma y$ [23 + 21] is equal to $\Sigma(x + y)$, and $\Sigma x - \Sigma y$ [23–21] is equal to $\Sigma(x - y)$!

In general, then, when you are summing a complex expression made up of terms, the result is the same as summing the individual terms and then combining these sums. However, you can't sum *within* a term (remember that terms are portions of the expression that are separated by plus or minus signs), and then perform a multiplication or division on those sums. (As noted earlier the contributions of the numbers to the sums are independent of order for additions and subtractions, but dependent for multiplications and divisions.)

To illustrate further, take $\Sigma(x + y - z)$; this equals $\Sigma x + \Sigma y - \Sigma z$. But the expression $\Sigma x(y + z) = \Sigma[x(y + z)] = \Sigma(xy + xz) = \Sigma xy + \Sigma xz$! The xys and the xzs have to be multiplied first, and then summed across the population. [Actually, this expression would better be written as $\Sigma[x(y + z)]$ to avoid confusion.]

One final point with respect to summations has to do with constants. Suppose we have a summation wherein the thing being summed includes a constant. Then what? If we take Σkx, where k is a constant attached to each of our x observations across the population, we can see that this consists of a series such as $kx_1 + kx_2 + kx_3 + kx_4 + \cdots$, where the subscripts refer to the different individuals in our study population. Because it is evident that each of the individual observations is k times as big as it would be without the constant, it follows that their sum will also be k times as big.

Think about this is though it were the reverse of expanding a parenthetical expression. If we had $k(x + y + z)$, this would equal $kx + ky + kz$, right? Then, of course $kx + ky + kz = k(x + y + z)$! So, the effect of summing an expression in which there is a constant as a multiplier is to move the constant outside of the sum (apply it to the sum, rather than to the individual values). Therefore, $\Sigma kx = k\Sigma x$. Obviously the same applies to a constant used as a divisor; for example, $\Sigma(1/k)x = (1/k)\Sigma x$, for the same reasons.

What about an additive or a subtractive constant? That is, what about a constant that stands as a term by itself rather than as a factor times a variable. Take $\Sigma(x + k)$, for example. Applying our procedure for summing terms, we have $\Sigma(x + k) = \Sigma x + \Sigma k$; but, since k is a constant (has the same value for each member of the population) and since the Σ means that we add up that value, whatever it is, once for each member of the population, Σk takes on the value of k times the number of individuals in the study population, or kN.

If there were a dozen persons in the study, and each of them had a five-dollar bill ($k = 5$), adding up the dozen people would give you 12 5s (k times the number of people), or 60 as the value for Σk. The case for subtraction is entirely analogous, Σk where k is a negative would give you a negative sum equal to the negative value times the number of individuals concerned.

VI. Summary

In this chapter, I have stressed the point that numbers are symbols that represent concepts; they are one of the two great partially overlapping symbol systems that distinguish man, language and numbers. I have noted that the nature of statistical applications is the linking of numerical concepts with the conceptualization of problems and issues of interest in real-world terms.

We reviewed and defined the major divisions of the number system. Statistics deals with "real" numbers, which are divided into the "positive" and "negative" numbers. Each of these is divided into "rational" numbers, comprised of "integers" and "fractions," and "irrational" numbers. There is an infinite number of these numbers which can be represented as a continuous line ranging from zero to infinity in either direction, with the positive numbers on the right and the negative numbers on the left.

We also discussed another type of number, called a "literal number," in which a letter is used to stand for some unspecified quantity. Literal numbers enable us to state general relationships among variables (things that vary from individual to individual). They do not differ in principle from regular numbers in terms of arithmetic procedure.

Finally, the role of estimation was discussed, and several areas of arithmetic operations of special concern to statisticians were reviewed: square rooting, spurious accuracy and rounding, and summation. It was suggested that all arithmetic/statistical results, including square roots, should be estimated as a guard against error. Rules of summation were developed and examples of their application were given.

VII. References

Drooyan, Irving and Wootin, William. *Elementary Algebra*. New York: Wiley, 1980.

Durren, J. H. *Statistics and Probability*. London: Cambridge University Press, 1970.

Forster, Alan, et al. *Algebra I*, also *Algebra II*. Lake Forest, IL: Macmillan (Merrill/Glencoe), 1992.

Keedy, Mervin L. and Bittinger, Marvin L. *Introductory Algebra, 2nd Ed*. Reading, MA: Addison-Wesley, 1986.

Kolman, Bernard. *Introduction to Linear Algebra with Applications, 5th Ed*. Lake Forest, IL: Macmillan, 1993.

Weil, Andre. *Number Theory for Beginners*. New York: Springer-Verlag, 1979.

2

Introduction to Statistics
and Data

29

I. Overview

This chapter begins by defining "statistics" and offering a number of observations and examples designed to provide a glimpse of the wide range of uses and applications of statistical ideas. I also note that statistical ideas rarely apply to individuals, but rather to the process of gleaning information and trends from large masses of data.

There are several overarching notions that have an important bearing on things statistical generally. These include the concepts of average, approximation, probability, randomness, and individual differences, and each of these is briefly discussed.

Statistics did not spring from a void. Nor does it exist in a vacuum. In this chapter, I lay a foundation for the later discussion of "things more statistical" by trying to set a context for statistics. Because the basic raw material of statistics is data—usually in the form of numbers, we must talk a bit about where such numbers might come from—some of the fundamental characteristics of measurement and the collection and/or assembly of the data on which the statistician operates.

Next, the point is made that, in one way or another, statistical notions tend to find their applications in connection with matters of scientific inquiry. [The term scientific should be thought of broadly here; perhaps systematic might do as well.] As statistics is thus frequently a tool in the service of science and research, I have digressed to summarize the nature of the scientific method and its most important technique, the experiment. In a word, these activities are the source of much of the data that we treat statistically.

Finally, although formal applications of statistics tend to take place in connection with systematic inquiries and research, every one of us deals daily with matters that are fundamentally statistical in nature. Thus, at some level we are all statisticians in our daily lives, and therefore have a vested interest in acquiring a better understanding of things statistical.

This chapter sets an important context for the more traditionally statistical material in the chapters to come. It is not difficult, and perhaps some of the material may be new to you. But, even if you find it desirable to skip and scan, let me first stress that statistics is not an old and moldy packet of esoterica. At some level it is an integral and vital part of all of that portion of human enterprise that is concerned with assembling, assimilating, and applying bits of information to the business of life.

II. What Are Statistics?

All of the previous palaver has been a precursor to talking about statistics because whatever statistics is (or are), it (they) come(s) in the form of numbers.

A. Three Definitions of Statistics

So, just what are (is) statistics? Well, there are three meanings of the term in common usage:

1. Statistics is a discipline

That is, it is a body of knowledge, rules, principles, techniques, and practices—like biology or astronomy or chemistry (examples from the scientific rather than the humanistic side of things, because statistics tends to be allied with science). Thus, you can study "statistics" and be a "statistician" in the same sense as studying biology and being a biologist, or as studying physics and being a physicist.

2. Statistics are data

That is, statistics are numbers that result from the application of the procedures of the discipline of statistics. Thus, they can result from collecting data, processing data, and reporting data. For example, we talk about Census statistics (meaning the data collected, processed, or reported by the Government based on the decennial census); or about economic statistics (meaning data collected, processed, or reported about the progress of various economies); or education statistics (meaning data about how students have performed on tests, or dropout rates, or expenditures per pupil, and so forth.)

3. A statistic may be the value of a particular index

A statistic may be the result of a particular statistical procedure. For example, if statistical data are processed so that an "average" is produced, the average itself may be called a statistic. In this sense, the word is close to the data meaning above, but is specifically the result of applying a specific statistical procedure or technique, rather than the body of data upon which the procedure was performed.

B. Statistical Usage

In common usage, the three meanings of the term, as outlined above, tend to get blurred, and it is sometimes important to examine the context in which the usage occurs to ascertain the exact sense in which the term statistic(s) is being employed.

It might also be useful to consider what statistics (the discipline) is for (and not for).

1. General considerations

The most basic purpose of statistics is to assist the analyst in bringing order and information out of *large masses of data*. Statistical techniques are designed to order or process large amounts of information in ways that bring to light trends, highlights, and summary features of the information.

The statistics of small groups (small amounts of data) becomes increasingly tenuous as the groups become smaller. Statistics does NOT apply to individual observations, and RARELY to individual persons (only when many observations have been made of the same person). In general, the larger the amount of data to be understood, the more vital is the use of statistics.

For example, if one has picked a dozen apples, and is curious about the weight of the apples being produced, one can weigh them, write the results on a piece of paper, and get a good idea unaided by statistics. For 120 apples, the procedure results in 120 numbers—still perhaps interpretable, but unwieldy. For 1200, or 12,000 it is impossible for the human mind to hold all of these numbers in mind and apprehend what information they hold regarding the weight of the apples being produced. (Note that for a single apple, no processing of numbers is required, and thus any statistical procedure is inappropriate.)

In the next page or two I am going to describe three major uses of statistics. However, I should like to emphasize that these three do not necessarily, or even frequently, exist independent of each other. Indeed, most practical problems involve the use of techniques from more than one of the categories discussed below.

2. Descriptive statistics

Within this broad context then, what are statistics used for? The first function is *description*. If we have a huge container with 1200 apples in it and someone says "Tell me all about your apple crop," statistical procedures can be applied to measurements of the size, weight, color, and other characteristics of the apples to produce understandable summaries of the data. Note that the issue here is simply to talk about what is there.

Such descriptive techniques, the particulars of which we will discuss in later chapters, can be applied to any large body of numbers (data), such as annual income of pediatricians, graduation rates from high school, farm production, ozone production, or whatever. The issue is simply that the sheer size of the body of data available boggles the mind without some set of procedures to abstract the essential information therefrom.

3. Inferential statistics

A second major usage of statistics and statistical procedures is for making *inferential* decisions. In the absence of complete data about some set of things, items, people, or whatever, it is sometimes useful or necessary to make judgments about the characteristics of the complete set based on the characteristics of only a few.

In order to make some statement about the yield per tree of apples in Washington State, for example, it is not necessary to collect yield statistics (data) for every tree in the state. Statistics (discipline) may be used to *infer* yield per tree in the state, based on the actual measurement of only a few, well-chosen trees. Similar sorts of procedures may be applied to the problem of inferring whether or not trees in Washington State yield more than do trees in Virginia. The use of statistics for inferential decision-making is also discussed in a later chapter.

4. Relationship statistics

The final major use of statistics that we will consider here is to describe the *relationships* among things. It has sometimes been said that everything is related to everything else in this world. Although this may not be literally true, sometimes it seems close! But the question is *how* related is it? Does smoking have any relationship to cancer? If so, how much? Does cholesterol have any relationship to heart disease? If so, how much? Just these two questions, both currently of wide interest in our society, illustrate the importance of the many issues that can be addressed under this relationship aspect of statistics.

Some more pedestrian questions might be: Do tall people, on the average, also tend to be heavy? Do people who do well on the College Board tests also tend to do well in college? In life successes? Do students who drop out of high school tend to make less money at the age of 25? Do fat people really tend to have a sunny disposition? Etc.

Note that in each case the answers are not either yes or no, and therefore are not obvious. That is to say, though the implied relationship may exist, it may exist only to a limited degree. Therefore, the statistical procedures that have been developed are designed to show not just that there is some relationship hidden in the data, but how strong it is, and whether it is "direct" (the more of this, the more of that), or "inverse" (the more of this, the less of that). Again, further discussion of this aspect of statistics will be reserved for a later chapter.

III. Some Important Notions in Statistics

There are of course many important notations in the statistical arena, and we cannot possibly discuss them all. Some will be reserved for the more detailed treatments offered in later chapters, and some perforce must be

omitted entirely, given the limited scope of this book. A few are sufficiently overarching to the areas we are treating that they are mentioned
here.

A. Average

Average is a very important idea in statistics. Chapter 4 makes clear that
there are many kinds of averages, but that can wait. What is to be stressed
here is that many important statistical techniques are founded in some way
on the concept of averages. This is also fortunate, because most people
have acquired this concept, at least at the common language level, and this
familiarity will make the understanding and interpretation of some statistics considerably easier.

The concept of average has to do with the idea of being like a lot of
others with respect to the thing under discussion—not being at an extreme. We speak of average height, average income, average intelligence,
an average year's grain production, an average summer's rainfall, an
average year on the stock market, an average day at work, average daily
production rates for the month, and so forth. In each case the critical
attributes may differ at the detailed level, but the basic concept of average
always entails middle-of-the-road, not-given-to-extreme ideas.

You will remember that there were three main areas of statistics to be
covered: descriptive, inferential, and relational. It turns out that statistical
techniques in each of these areas depend heavily on the idea of average.
In descriptive situations one of the things we most frequently offer in
describing a set of data is an average, and furthermore, at least one other
characteristic of data that is commonly described (variability) is dependent
upon the idea of average for its interpretation.

In inferential situations, the inferences to be examined are often about
averages used to describe one or more sets of data, and to some extent the
examination itself is based on underlying applications of the idea of
averages. Finally, the major index used to describe the relationship between two or more sets of data can be expressed as an average, when
reduced to an unfamiliar, but basic form. Certainly, then, the concept of
average is one of the most important ideas in all of statistics.

D. Approximation

Another fundamental concept in statistics is that of "approximation." It
has been said (only half in jest) that if it is a statistical fact, it is WRONG.
This is because the most common situation in statistics is that the conclusions to be drawn are based upon less than the entire set of items or

events of interest. (This would have been the case with the Washington State apple tree yield mentioned earlier.)

In cases where there are large amounts of data, and the use of statistical techniques is thus appropriate, it is exceedingly common to base studies and conclusions on relatively small subsets of the data. In such situations the results cannot possibly be absolutely precise, as compared to those that would have been obtained if *all* items in the potential data set had been examined. With the single exception of those instances where all of the potential data points have been examined, statistical studies always result in approximations.

[It should be noted that approximations are not bad in and of themselves, but this raises the issue of whether they are close enough to be useful for the purpose at hand—or, "is it good enough for Government work?" Those who use statistical results should understand that, no matter how well-designed the study was: (1) the results are not numerically precise; and (2) it is necessary to seek information as to how precise they really are.]

If the usual procedures yield approximations, why don't statisticians always deal with the entire population of interest? There are at least four reasons.

1. Complete data are not necessary

First, in many cases it isn't necessary to deal with the entire population, because *acceptably precise* results for the purposes of the study may be attained by looking at only a fraction of them. This is particularly true where it can be established that the items of interest are relatively alike.

For example, if a factory producing tennis balls wants to examine the weight of the balls coming from a given production machine, it is not necessary to weight every ball to get a good idea of how much the balls produced by this machine weigh. If the balls differ by only 1 part in a million, not even Steffi Graf will care, and a sample of a few will be sufficient to establish accurately what the production weight of these balls is.

(Note, however, that the more diverse are the items in a set, the larger proportion of the set that must be examined in order to be able to accurately describe the set. When diversity is extreme, it may well be that this particular set should be reexamined at the conceptual level to ascertain that it is a suitable set for study. Further examination will frequently reveal that there were important inconsistencies in the way the set of items was defined in the first place which have resulted in mixing together "apples and oranges." It may be better to examine such a set by redefining it into subsets according to some other critical attributes, so that the subsets are more homogeneous and amenable to study.)

2. Complete data are costly

Second, dealing with the entire population of interest is frequently intolerably time-consuming or expensive, or both. For example, educators frequently like to administer "essay" (free response) tests to their students. Many have asked that such tests be used instead of, or in addition to, multiple-choice tests in the process of college admissions.

Although some of this is now being done, colleges and testing services have maintained that with millions of applicants, widespread use of such procedures would require staff and money to administer and score that would be far beyond their resources, even if the student burden of millions of hours of testing time could be accepted. Thus, these kinds of data are generally not available because of cost, requirements, and time. As a consequence, statistical studies of essay data are of necessity based upon smaller groups of students than the total college admissions population.

3. Some studies are destructive

Third, some studies are by their nature destructive to the items under study. If this were the case, one would certainly prefer not to subject all of the items to the procedure just in order to have complete data.

Take, for example, the California Condor. At this writing there were some 28 left in the world, and all were in captivity. Authorities wished to study the advisability of releasing them into the wild. They planned to try it with only four. Obviously if all 28 were released, complete data would be available; but if the experiment should prove fatal to those released, there would be no need to study the California Condor any more—the approximation provided by the group of four would have been much to be preferred. Tests of bullets, bombs, balloons, soap bubbles, and many other things are of this ilk.

4. Complete data are impractical

Finally, there are situations where the set of potential items to be studied is either infinite or practically infinite, so that it is not possible to test all items. An example of the former would be tossing coins. If you want to measure the percentage of tails produced by tossing a coin or coins, there is theoretically no limit to the number of tosses you can make—the limit is only how long you are willing to keep at it—millions and billions of tosses —if you live that long!

An example of practical infinity might be an attempt to measure the body weights of persons who live in America. Theoretically, this is a finite group. However, practically, it would take a lot of time, and even if it were possible to do the measurement instantaneously across the country at the same time, it would be wrong the next instant because during that time some people would eat a big pizza, some would die, others would be born, and thus the numbers would change.

These kinds of considerations are seen as a real problem by the Census Bureau, which is required by law to conduct a complete count of the United States population. There have been suggestions to substitute sampling processes from time to time, but these do not seem to meet the constitutional requirement. In dealing with this problem, the Census Bureau tries to collect its data *as of a particular day* (such as April 1 of the censal year), and uses sampling techniques extensively in its supplementary studies.

C. Probability

Another idea that has found wide application in statistics is the concept of probability. This grows out of the necessity for approximation as described above. If one can't usually obtain precise results, the next most comforting thing is to be able to state a probability concerning the correctness of one's approximations. Thus, many statistical findings and results have a statement of probability associated with them that is intended to convey some idea of the goodness of the approximation given.

I will make no attempt to cover the details of probability theory here. The interested reader is referred to the reference list. However, at its most basic, probability is a numerical statement of the likelihood of getting a certain result from a procedure under a given set of conditions.

Probability is expressed as a numerical fraction ranging from 0 to $+1$. (We do not deal with negative probability.) The common expression that is usually associated with probability is, "What are the chances of...happening?" Odds, as in horse racing are a related concept, but, since statistics are rarely phrased in terms of odds, we will stick to "What are the chances of...happening?"

The general rule for calculating the probability of getting an outcome with a given characteristic is to divide the number of possible outcomes with the given characteristic by the number of all possible outcomes.

Suppose you had 9 marbles in a box, and they were all white ones. If you drew out 1, without looking, the probability of drawing a white one would have been 9 (the number of possible outcomes fulfilling the characteristic of white) divided by 9 (the total number of possible outcomes). $9/9 = 1$. A probability of 1 (sometimes referred to as 100%) is called "certainty"; it means that the result is inevitable—no exceptions. On the other hand, the probability of drawing a red one from this same box would have been $0/9$, or 0. No chance! Zero means it will never happen.

Now suppose that someone adds 3 red marbles to our box. Now, given a single blind draw, there are still 9 outcomes with a white marble, but the number of possible outcomes of a single draw is 12. Accordingly, the probability of drawing a white marble on a single draw is $9/12$, or 0.75. If, on the other hand, the person who put the red marbles in took out 3 white ones to keep the total number constant, there would be only 6 ways to get a white one out of 9 possibilities, and the probability would be $6/9$ or 67%. (It isn't really correct to express probability as percentages, but it is frequently done—simply by multiplying the probability fraction by 100.)

Taking one more example, suppose that someone comes up to the original box of 9 white marbles and *adds* 3 *marbles, all of the same color*, but without telling us what color they are. Now, when we blindly draw a single marble, and we get a white one, what can we say? Only that the probability that the 3 new marbles were in fact also white is $3/12$ (the probability of drawing 1 of the 3). This is because if the 3 new ones were the same color as the original 9, there would be no way of knowing for sure whether we had drawn new or old, and no definitive statement about the color of the new marbles would be possible.

On the other hand, if we had by chance taken one of the 3 new ones, AND, if it were *not* white, we would have demonstrated that the 3 new marbles were of that new color, and not white.

In any case, we could state a *probability of having new information* as being the probability of drawing one of the new marbles, not the old, or $3/12$, which would also be the probability that the three new marbles were whatever color we drew. WHEW!

Now, let's think a little more about how all of this may apply to statistics. First, it is important to understand that we cannot demonstrate or prove anything about the three added marbles UNLESS we just happen to draw one, AND, we know that the new marbles in the box were of a different color. But, in most real statistical applications, the exact distribution of the characteristic in the population of interest is unknown to us: frequently, we are doing the study to find this out.

This is the important thing: Statistics and probability usually don't prove things beyond doubt; they only permit approximations, or estimates, sometimes with stated probabilities of correctness.

Second, probabilities are used in conjunction with some kind of statement or conclusion to approximate the chances of being wrong. For example, if I estimated the apple tree yield in Washington State to be

20 bushels, based on a small subset of all of the trees in the state, I might attempt to qualify this estimate somewhat as follows: "According to my calculations, my estimate (20 bushels) won't be off the true figure (if we only knew it) by more than 2 bushels either way—19 out of 20 times (probability of 0.95).

In other words, 20 bushels is an estimate or approximation; but, I am confident that, if my study were repeated a large number of times, 95 times out of 100 I won't have missed the figure I would have gotten if I had looked at all of the trees by more than 2 bushels either way. Or, to put it another way, by combining my calculated estimate with probability considerations, I have created a *range* within which I could expect to find the unknown true value 95% of the time. (Please note that these yield figures are just so much applesauce, as I haven't done the study to find out how much a Washington apple tree really produces.)

Probabilities are used in some other ways in statistics as well. We will take a more detailed look at the way probability statements are used to give meaning to statistical estimates in Chapter 8, Statistical Inference. Note, again, that there is no statistical way to *prove* the exact number of bushels, short of looking at all of the trees in the state—a procedure that would very likely be too costly and time-consuming to be a practical alternative.

D. Randomness

Many statistical procedures depend on selecting a subset of items, people, or things from a larger set "at random." Technically, the requirement for such a selection is that at any given draw, the probability of choosing a particular item from the larger set is exactly the same as for any other particular item. (We'll go into such "random sampling" in more detail in Chapter 9.)

Thus, if there were 129 items, the probability of selecting any one of them would be $1/129$ (the number of ways the designing outcome could occur divided by the total number of possible outcomes). Loosely, we would say those picked would be picked strictly by chance—meaning that the chances (probability) of picking any particular one were equal.

It is axiomatic that randomness is best ensured by using a mechanical rather than a human agency to do the picking, because humans frequently have biases and preferences that even they don't know about. Machines, if properly designed, can't care.

There are a number of sources to assist us in getting random numbers. Tables of random numbers exist (produced by machines) and computers often have random number generators. Other mechanical devices exist to produce random results, as well.

E. Individual Differences

A final notion that underlies the use of statistics is that of "individual differences." This idea comes into play when we realize that the normal condition among a series of measurements (of almost any kind) is that they differ among each other. A familiar example is fingerprints. We have all heard that no two persons have exactly the same fingerprints, and indeed, that law enforcement and the courts recognize this fact. So, too, we hear that no two snowflakes are the same.

Correspondingly, this condition is so well recognized in statistics that the dimensions on which we take the measurements for statistical analysis are usually called "variables." This means that we expect the measurement values to differ from one case to another.

Individual differences are important to statistics for two reasons. First, in any large mass of data we can expect variation—it won't be enough to simply look at a few observations. Statistical techniques will be required to tease out the relevant information.

Second, it means that a legitimate area for study will be the ways in which things vary—that is, the nature of the differences themselves. For example, when we manufactured tennis balls, we are interested not only that the balls have the correct weight, size, etc., but also in the pattern of deviations of the individual balls from the desired optimal values. This kind of investigation of individual differences is the basis for what is often called in manufacturing "quality control"—an analysis designed to measure and control the variability in the production process.

As a consequence, statistics has developed a series of techniques to study such issues of variability.

IV. Some Measurement Notions

So far, we have dealt largely with the idea that statistics is concerned with numbers, or data, and the basic definitions and procedures for manipulating numbers. Now, it is time to consider the fact that in real life the numbers processed in statistical ways generally come from some kind of a measurement process.

Obviously, the source of the numbers that are to become grist for the statistical mills is important. It is difficult to choose the appropriate techniques, and to make the appropriate interpretations of statistical results, in ignorance of the nature and characteristics of the measurements underlying the numbers to be processed. Therefore, the next few pages are devoted to the discussion of some notions about measurement.

A. What is Measurement?

Measurement is the process of attaching judgments, usually in numerical form, to observable aspects of things or events.

There are several points worth noting about this definition. First, the process is always judgmental at some underlying level, even though it may look mechanical. For example, when you take the written test for your driver's license, it may be scored mechanically, and there may be an established passing score. Nonetheless, someone's judgment was used to:

- Select the topics to be included in the test
- Choose and phrase the questions
- Choose and phrase the correct answers
- Choose and phrase the incorrect answers or "misleads"
- Determine how much each question should count toward the score
- Determine the size and nature of a passing score.

In some measurement situations there is also an element of judgment inherent in the interpretations that are made of the direct results of the measurement procedure. For example, a college Admissions Officer may take several measurements into account in making a judgmental decision to admit or deny admission to a prospective student. In various measurement situations it is not always obvious as to whose judgment it is, or where it enters the picture, but you may rest assured that a copious quantity of judgment is in there somewhere.

For measurement to take place, it is not necessary to have an explicit or formal scale or test. However, the event or situation must be *observable*. But, you say, what about situations such as an interview for a job, or for college entrance? Yes, the interview is a measurement situation, and may result in a measure such as a numerical rating, or a nonnumerical outcome such as approval or disapproval. The important point is to recognize that whatever the process was called, the measurement results from the interviewer's reactions to what he or she *observed*—your overt behavior, or your test responses, or whatever other data you provided in response to a defined (or even ill-defined) situation.

You cannot measure something that is unobservable (invisible to whatever instruments or procedures you are trying to apply). For example, ideas, concepts, and abstractions cannot be directly measured because they are unobservable; only from inferences about their presumed, or logical, relationships with observable events and data can we draw conclusions about them.

Frequently, however, we wish to measure things that are not themselves directly observable, such as leadership, job fitness, intelligence, or personality. In order to do so, that is, in order to engage in measurement at all, we must translate such concepts into something that *can* be observed—an "operational definition." Operational definitions are the *observable* behaviors or events that *can* be measured, and that provide us with the basis for inferring the status of their underlying concepts. Obviously, the logic that establishes the linkage between the observable behavior measured and the presumed underlying concept is critical!

I have noted that measurements may be either numerical or nonnumerical. Intuitively, we would favor numerical measurements, simply because the process of measurement is usually directed toward gathering information, which in the normal course of events, tends to be more valuable if it has the precision imparted by numbers. So, from here on, I'll talk about *numerical* measurement.

The first requirement is to decide what it is that we want to know about. This entails a very careful choice of concept to be studied or measured. It is necessary to develop a complete understanding of and definition of this concept in order to assess its measurability; to be able to develop and justify an operational definition if, as is quite likely, one is needed; and to be able to interpret the results of the measurement accurately once they are obtained. This step forms the basis for deciding just what is to be measured.

Once one has decided and defined what is to be measured—crop yield, educational achievement, or whatever—the next decision is how to do it. There are many particular devices and techniques that may be used, such as tests of various types, different sorts of rating schemes, schedules, work samples, time-sampled observations, surveys, etc. We will not try to review them here. The reader is referred to a text on psychometrics (measurement) for details (see the reference list). What is of interest to us, however, is that in the sort of measurement we are talking about, *these devices and techniques result in a procedure for assigning a number to indicate the frequency of or the amount of the thing being measured.*

1. Units and scales

In the process of measurement, the value assigned always has a "unit" attached to it. That is, it is not a number such as 17, or 1000; it is 17 somethings, or 1000 somethings. The choice of a unit for the measurement sometimes arises from the nature of the concept to be measured, for example, if you are counting "things," then "1 thing" is the unit.

Sometimes it arises from the operational definition for example, if you operationally define "aggression" in terms of a child striking another child, then "1 hit" becomes the unit of aggression. Sometimes the unit is inherent in the measurement technique, for example, if it is an achievement test, then the unit usually becomes "1 question answered correctly."

In numerical measurement, the concept is usually thought of as being a dimension or trait that exists in varying amounts among the individual cases or items to be measured (in other words, a variable). Theoretically the number of units that might result from a measurement can be thought of as being arranged from none or a few (representing little of the concept) in increasing numbers up to all or many (representing much of the concept).

When an arrangement comprised of all possible numerical results of the measurement is laid out in terms of these units, from minimal (usually on the left) up to maximal (usually on the right), it is called a "scale" (see below). Thus, the number of units resulting from the measurement of an individual can be referred to the scale in order to evaluate how much of the concept was present in that individual.

Little   Intermediate Values   Lots

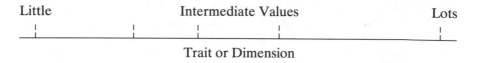

Trait or Dimension

2. Continuous and discrete variables

The theoretical measurement dimension (underlying concept) can be thought of as being either "continuous" or "discrete." A continuous dimension is one in which the logical analysis of the concept suggests that theoretically there are no gaps in the possible amounts of the dimension that might exist. For example, if one were to measure the amount of water in various lakes in the world, or the amount of intelligence of various people, there would be no reason to suppose that any particular values within the scale would be nonexistent. If such a scale were to be represented graphically, it would be a continuous line from low to high, showing that all intervening values were possible (as above).

If, on the other hand, there should be logical gaps in the dimension, the scale based on it is called "discrete." The most obvious examples of discrete scales are those that represent counts of things. We all chuckle when the Census Bureau reports things such as "the average U.S. family is comprised of Mom, Dad, and 2 1/2 children." We know that fractional children don't exist. That is, the scale representing number of children in the family is discrete—it exists only in terms of whole numbers with no intermediate values between. If such scales were represented graphically, they would consist of a series of dots, arranged in order from low to high, and representing the whole-numbered values on the scale.

The difference between discrete and continuous theoretical dimensions is of course important in and of itself. However, it is more important to recognize that *the operational measurement scale is ALWAYS discrete, regardless of the nature of the dimension. This is because the scale must be defined in terms of units, and* operationally, *the measurement doesn't exist between units.*

You see, the concept of unit usually implies the idea of "indivisible amount." This is not always strictly true, of course. We sometimes report things in terms of fractions of units where this is not logically inconsistent with the concept being measured. But, it is *usually* true because the units were chosen to be at or very near to the smallest amount of the concept that could be meaningfully measured.

An ordinary ruler is usually divided into 16ths of an inch. Operationally, this is the unit being used. It doesn't mean that 32nds are not meaningful in a theoretical sense. But most people cannot use a ruler with units smaller than 16ths dependably. So they rarely use or report measurements based upon estimating the distance between the 16th marks on the ruler. If a finer precision is required for a task, a different sort instrument is used for the measurement—one with a vernier scale perhaps.

Or, take an intelligence quotient (IQ), for example, we might say that someone has an IQ of 108, but it would be unusual to say 108.3, and unheard of to say 108.347.

Note that even, if we ignore the issue of the appropriate units to use, all measurement is still discrete. If 108.347 *were* meaningful, it represents a continuous scale divided into 1/1000ths of a whole unit. Practically, such a scale implies that values in between 1/1000 and 2/1000 don't exist. Of course, *theoretically*, they do. This argument can be extended to as many decimal places as you wish.

B. Functional Numbers — Kinds and Properties of Numbers Used in Measurement

In the previous chapter we reviewed numbers classified from the point of view of the mathematician. Now, in looking at the sources of numbers that find their way into statistical procedures, we need to review some of the ways that they may be classified according to their functional properties. This is the form in which they are used in measurements.

The application of numbers and number manipulations to measurement situations must be done with considerable care. Not only must we follow sound procedures for defining concepts with critical attributes, develop measurement dimensions and operational definitions of that which we wish to measure, construct valid and reliable measures, collect the data appropriately, and manipulate and combine numbers accurately, but we must also assure ourselves that the types of numbers used, in the particular applications in question, make sense and are in accord with the kinds of meanings to be expected.

As you will see from the discussion below, in the functional sense each of the different kinds of numbers implies different sorts of measurement operations.

1. Pure numbers

There are of course "pure" numbers. These are defined as the numbers *without* any units attached. Obviously they are not numbers arising directly from measurements, which always carry units. They are the numbers used in abstract discussions about numbers. They also result whenever you take the ratio of two measurements, both expressed in terms of the same units, which gives you a proportionality expressed as a pure number.

For example, it has been said that a good martini requires gin and white vermouth in the proportion of 3 to 1. The proportion is independent of the units, or "pure," since it could be 3 ounces to 1 ounce or 3 gallons to 1 gallon. Sometimes such ratios result in pure number constants, such as the constant pi to which we referred earlier. Pi is defined as the ratio of the circumference of a circle (in units of length) divided by its diameter (in the same units).

However, as noted above, in statistics we are almost always concerned with functional numbers, the ones used in counting things, the ones to which a unit is attached. These numbers, like "all Gaul," may be divided into three parts: nominals, ordinals, and cardinals. Each of these three kinds of numbers has different applications and uses.

2. Nominal numbers

The simplest of the number families is the nominal group. The word nominal derives from the Latin for "name," and this reveals the basic nature of these numbers. They are numbers used for the purpose of naming.

Suppose you are giving a children's party, and you have divided those attending into three groups. You intend to get one group busy playing a game while you take another outside and the third is served refreshments. You start out by saying, "Now, you children come over here and be Group

1, and you (other) children, you'll be Group 2, and the rest of you are 3. Now, Group 1, here's what you're going to do..."

Notice that in this example there is still very little feeling of numberness associated with the use of 1, 2, and 3. They are really used as assigned *names* for the groups for the convenience of having something to call each group when speaking to them or referring to them. In reality, you might just as well have called the groups A, B, and C, or George's, Sally's, and Mike's (after one of the children in each), or something totally esoteric, such as Bingle, Argle, and Mornswallow (sounds like a law firm!). Of course, there is almost always a little bit of numberness involved—you probably will give the group you called 1 its instructions first, for instance. But, clearly, the naming function is the primary usage of the numbers in this case.

a. Permissible operations with nominal numbers. What operations can we perform with nominal numbers? Well, it should be easy to see from the above example that it is meaningless to add, subtract, order, or otherwise manipulate these numbers. Their only meaning is membership in a group. Note, however, that membership in a group is an important quality. If Group 1 consists of my business enemies, Group 2 of my business supporters, and Group 3 is neutral, it behooves me to be able to distinguish the members of one from the members of another.

Nominal numbers, because they do not possess the property of quantity, but only name, can be used only in measurement operations wherein the primary problem is to sort, classify, or group objects. The measurement operation involved is simply to ascertain the presence or absence of whatever the concept in question is, so that the object may be classified regarding that concept or measurement attribute. Categorization or classification implies mastery of the critical attributes of the concepts(s) that underlie the formation of the groups, and that are the basis for their differentiation as groups and their homogeneity within groups.

In sum, then, nominal numbers lack the number qualities we usually think of, but are used as names to represent the different results of a classification process which may be either simple or complex in nature.

3. Ordinal numbers

A more complex, and more useful, kind of number is the ordinal number. This number takes its name from the word "order." As you might assume, these numbers are representative of position with respect to some well-defined characteristic.

For example, at high school graduation most high schools tell each student what his/her position was relative to the other class members with respect to grade-point average. This is known as "class rank," and is an important bit of information for those who aspire to admission to an institution of higher education. Thus, if Susie is Rank 12, she knows precisely that 11 other students had higher grade-point averages than she; if Charlie is 95 and there were only 96 people in the class, he knows (or should) that there was only one person more hapless than he.

Ordinal numbers are derived by arranging the available items from top to bottom or bottom to top with respect to the characteristic in question, and then assigning numbers to each item, consecutively, beginning at 1. Obviously, the same information is contained in the set of numbers whether you begin at the top or the bottom, but for interpretation you must know which was used as 1. Succinctly then, an ordinal number reveals the position occupied by that item among all of the items when they are placed in order according to a particular, defined characteristic.

a. Permissible operations with ordinal numbers. What operations are permissible with ordinal numbers? Fundamentally, the only new operation that has been added, beyond nominal numbers, is the operation of *ranking*: this term means only that the items in question are *arranged in order* with respect to some defined characteristic. [Of course, everything that was permissible with nominal numbers—which wasn't much (just naming) is also permissible.] It is still meaningless to add, subtract, multiply, and so forth. Note that adding positions 3 and 4 yields 7, while adding positions 6 and 1 also yields 7. But it should be plain that these two sevens don't necessarily represent the same amount of whatever it was that the position arrangement was based on.

However, this is an important step forward, as it permits the mathematical statements of "greater than" and "lesser than." We can therefore say that if ranks run from top to bottom, the item with rank 16 has more of whatever the ranking characteristic was than did the item with rank 20–and that both have more of it than rank 30.

Let's take a brief example. Suppose you were a basketball coach, and you had four applicants for center on your team, which you ranked by height. It might look like this:

Name	Height	Rank
Pete	6'10"	1
George	6'7"	2
Sam	6'6 1/2"	3
Schnikelfritz	6'6"	4

The point is to satisfy yourself that doing arithmetic on the ranks is fraudulent, false, and just plain unwise. For example, let's take addition. If you add ranks 1 and 3, you get 4. However, this would correspond to 6'10"+ 6' 1/2" or 13'4 1/2," whereas rank No. 4 is only 6'6."

You can easily see that any other arithmetic operation performed on ranks gives sheer nonsense with respect to the underlying dimension of height. It is very important to remember that this is *always* true, whether you happen to know the actual underlying measurement or not! The only statements that are safe are ones that aver that an item at given rank has more of the quality being ordered than those below it, and less than those above.

To summarize, the unit of an ordinal scale, a rank, is *not* a unit of the dimensions being measured, but merely a unit of *relative position* on the scale of that dimension. The units of rank are equal-appearing in size, that is, generally proceed by whole numbers (ranks 1, 2, 3, etc.), but in reality overlie varying and quite different size intervals in terms of the scale of the underlying dimension. That is, the difference in the amount of the trait represented by a difference of 1 in the rankings is an unknown, and is expected to be quite different at various points in the set of ranks. It is this characteristic of ordinal numbers, and rank measurements, that prohibits their use in most arithmetic operations, as most arithmetic operations demand numbers containing quantitative as well as relative information.

4. Cardinal numbers

It is only when we get to cardinal numbers that we are talking about numbers in the sense that most people usually think of numbers. As used here, the term "cardinal" means of prime or chief importance (no, it doesn't mean "for the birds!") Cardinal numbers are the numbers used to express actual counts, quantities, and amounts, not just names and ranks. Of course, as before, numbers capable of a given function can also be used for an lower (less precise) function. That is, numbers that are exact amounts can also be used to rank order items or to name items, as needed.

a. Permissible operations with cardinal numbers. The most immediate application of cardinal numbers has to do with what we all thought numbers were for in the first place—counting. We don't count with nominal numbers or ordinal numbers, although we might well count the number of items in the groups marked off by nominal categories. This is because counting is a precise use of numbers in the quantitative sense, as described above, and almost invariably employs a unit; that is, we count "somethings."

Counting (or, technically, "enumeration") consists of pairing off all of the items in a defined group or class with a sequence of natural numbers beginning at 1 and ending at the last member of the group. The number that is assigned to the last unit constitutes the count. A count is a precise quantity that may be subjected to various arithmetic operations.

So, with cardinal numbers, a 7 means exactly 7, no more and no less. Such precision permits the standard arithmetic operations such as addition, subtraction, multiplication, division, and so forth. When we know that we are using cardinal numbers, and we are told that George has 32 items and Pete 27, we can say that George is exactly 5 ahead of Pete; or we can say that George possesses 32/27 or 1.185 times as many as does Pete. If 32 and 27 had been ordinal numbers, however, the most we could have said would be that George exceeded Pete.

b. Some constraints and considerations. The above considerations seem quite straightforward. However, when we deal with cardinal numbers we must be very careful. We know now that cardinal numbers are those that imply the full numerical qualities of the number system. Thus, when we are presented with measurements in terms of cardinal numbers, we are likely to assume that they are fully meaningful. But, this is not always true, because they may be incorrectly or misleadingly employed in a measurement operation. Let's see how this could occur.

The use of cardinal numbers in their fullest, quantitative meaning in measurement implies that you can meaningfully perform all of the arithmetic operations we have previously discussed on these numbers. However, in order for this to be true, it further means that the numbers are attached to a scale with the following properties: equal-sized units and a true zero. We'll discuss these two properties in the ensuing paragraphs.

(1) Equal units. Suppose we have a scale that we will represent graphically as follows:

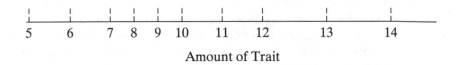

Amount of Trait

It is clear from looking at this scale that it does NOT have equal units. That is, 1 unit is not worth the same amount of the dimension at different points on the scale. Take a piece of paper or a ruler and prove to yourself that the linear difference between adjacent scale values, which represents the amount of the trait necessary to get from one scale point to the next, can vary quite a lot at different points on the scale. For instance, it would appear to take a lot more of whatever is being measured to get from 12 to 13 on the scale than it would to get from 9 to 10.

Thus, if we perform some arithmetic on values from this scale, we will get nonsense. For example, if we subtract 9 from 10, we won't have as much of the dimension as if we subtract 12 from 13, even though we get the same amount on the scale, 1! Now, this is very undesirable in a measurement problem, because it becomes impossible to compare various individuals, as the differences among them may be attributable merely to differences in the scale at different points.

It may come as a surprise, but percentage scales tend to have this difficulty. It is often the case that the percentages are attached to a dimension reflecting the distribution of some phenomenon in the population. For many phenomena it is true that most people are pretty similar and only a few are greatly different. Therefore, that amount of the phenomenon required to move ten (or one) percentage points in the distribution is far less at the middle than it is at the extremes of the distribution. See the example below. We'll talk more about this in connection with distributions in the next chapter.

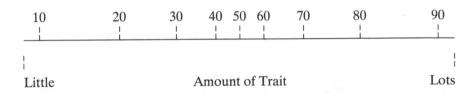

Overall, then, a good measurement scale should have units that are the same size at all points along the scale. Such scales are called equi-unit scales or "interval scales", because equal-appearing intervals on the scale represent equal intervals on the underlying dimension. This makes possible additive and subtractive operations, which means that we can compare the scores obtained; for example, Johnny has 45 and Pete has 36, and therefore Johnny is 9 points better than Pete! Furthermore, if Susie has 40 and Marge has 49, it is reasonable to say that Marge is about as much better than Susie as Johnny is better than Pete. Such comparisons cannot be made meaningfully unless the scale has the equi-unit property.

(2) True zeroes. A true zero is simply the point on a scale where there is exactly none of the phenomenon being measured. At first blush this doesn't seem too formidable. It is easy to think about exactly none of things such as money or counts of objects. However, in the social sciences, where statistics finds frequent application, many of the measurement dimensions are difficult to conceptualize at the zero point, and, indeed, impossible to measure there, even if such a point were to exist.

For example, it is easy to think of exactly no money or assets; it is more difficult to think of exactly no education; and it is impossible to conceive of exactly no intelligence (since long before you get to this point you no longer have a functioning human being).

So, what difference does it make? Well, without a known point on the scale corresponding to a true zero, some of our arithmetic operations cannot be meaningfully conducted, even if we are using cardinal numbers, and even if the measurement does have the equi-unit scale characteristic. When we talk about equi-unit scales, we are talking about scales for which addition and subtraction are legitimate at any point on the scale. However, unless we know the zero point of the scale, we cannot say that Johnny's score is 5/4 of Pete's, or 25% better. Taking a ratio (multiplying or dividing) requires *both* equal units *and* knowing the zero point of the scale. *Scales that have both the equi-unit property and the true zero property are called "ratio" scales.*

Let's take a simple example first:

	0	1	2	3	4	5	6	7	8	9		(assumed scale)
[0	1	2	3	4	5	6	7	8	9	10	11]	(real scale)

If the assumed scale is the one that we get from our measurement, we will be tempted to manipulate the numbers in their fullest cardinal sense, including multiplying and dividing. However, if the real scale (shown here in brackets, *but actually unknown to us*) has a different zero point, representing exactly none of the phenomenon measured, ratio comparisons made using the assumed scale will be inaccurate as compared to the corresponding comparisons made using the real scale. Unfortunately, this is usually the case!

Continuing to look at this example, if we say that a score of 8 means twice as much of the dimension as a score of 4, but in the real scale this represents a ratio of 10 to 6, it is obvious that the true comparison is 1 2/3 as much rather than 2 times as much. Again, 2 appears to be half of 4; but, if the true zero were known, we would find that 2/4 is really 4/6, or 2/3, which is not the same thing at all!

Another way to say it is that we want ratios that appear to be the same to actually be the same. In the assumed scale above, 3 is to 2 as 6 is to 4 (sometimes written symbolically, 3:2 :: 6:4); but in the true score scale these numbers would actually be 5:4 :: 8:6—which is clearly not so.

So we need to be able to fix the end-point or true zero in order to have a scale where equal-appearing ratios are indeed equal, that is, a ratio scale.

Let's work with a common scale from physics as another example—the temperature scale. If we use Fahrenheit, freezing is 32°, and boiling is 212°; suppose the current temperature is 16°F. Can we say that it is 16° below freezing? Of course. Can we say that it is 8° colder than it was yesterday, when it was 24°? Surely. This is because the *Fahrenheit scale as commonly used is an interval scale, and subject to additive and subtractive comparisons*.

Now, can we say that if the temperature today (16°F) doubles tomorrow it will be freezing? No, not at all. As a matter of fact, if it doubled, we'd all be fried! Sure 32 is twice 16—but this is with reference to an *artificial zero* (0°F), not to a temperature representing exactly no heat (the true, or "Absolute" zero). Therefore ratio comparisons may not be made with the common Fahrenheit scale.

Absolute Zero in the Fahrenheit scale is $-459.6°$. This is the point where molecular motion ceases and the heat thus generated falls to zero. Thus, with reference to the true zero point, 16°F is really 475.6°F, and double it is 951.2° "true" Fahrenheit (or 491.6°F according to our common scale—much too hot for comfort!)

The point here is that unless the zero point on the scale represents exactly none of the phenomena being measured, any multiplication or division operation will produce a misleading result. The temperature scale is an excellent example because the commonly used zero is certainly a long way removed from no heat. But, in contrast to most situations, in this case we actually know what the true value is, so that we can illustrate accurately the extreme difference between calculations based upon the common, false, zero and those based upon the true zero.

By far the more common situation is that we don't know the true zero for a scale. Unfortunately, sometimes this lack does not deter the statistician or the user of the measurement, who proceeds to operate on the numbers anyway, as though the scale were truly a ratio scale. This is the kind of thing I mean when I say that cardinal numbers in measurement cannot be applied blindly and without care, even though it might seem appropriate to do so at first blush.

C. The Reliability and Validity of Measurement

It would appear that in general there are many opportunities for distortion in taking a measurement. The concept may be poorly defined, the operational definition badly rationalized, the observations incorrect, the technique flawed, the scale inaccurate, the interpretation misleading, and so forth. This raises the need to accept measurements cautiously and, if possible, to assess their goodness.

There are two ways in which measurements are generally assessed. These are in terms of "reliability" and "validity." Again, the interested reader is referred to a psychometric test for details (see the reference list), but it is possible to give the flavor of these two concepts briefly; and, it is important too, since tests and measurements have assumed an increasingly important place in modern society.

1. Reliability

Reliability is conceived of as an assessment of the *accuracy* of the measurement. It does not deal with the issue of the relationship of the measure to the underlying concept or the adequacy of the operational definition. It merely asks the question, "Whatever it is that you are doing or why, are you doing it accurately?"

The reliability of a measure can be estimated in several ways. It is most fundamentally addressed by asking the question, "If a set of measurements could be repeated, independently of the first time, to what extent would the repeated measurements yield the same result?"

2. Validity

Validity, on the other hand, is concerned with *the extent to which the measure used truly measures what you thought it did (and wanted it to)*. In other words is this a "true" or "valid" measure of the concept being assessed?

This question is usually approached by comparing the results of the measurement to those of a different (criterion) measurement of the same concept, or to a set of logical deductions about the nature and content of the construct or concept being measured. Validity is more difficult to establish than reliability and is fundamentally the more important of the two. It is possible to have a reliable measure that has nothing to do with what you really wanted to measure in the first place. However, validity (measuring what you intended) implies reliability; you can't measure what you wanted to measure if the measurement is all wrong!

3. A Comparative illustration of reliability vs. validity

A simple example may help to understand these two ideas. Suppose I set out to measure intelligence. (Note that this is one of those concepts that cannot be directly observed.) I reason that intelligence is a function of brains, and brains are located inside the head, and therefore measuring the size of a person's head is the way to measure intelligence. I select use of a tapemeasure as my technique and proceed to measure the circumference of the head for each of the individuals in my study population.

Certainly I can do this quite accurately. I can even prove it by having someone else make the same measurements, which will compare quite nicely with my set. In other words, the technique has *reliability*.

However, upon further study, I can find no relationship between my measures and anything that is commonly understood to characterize intelligence. My "big-heads" don't do any better in school, don't hold better jobs, don't create more, and don't seem more literate. It would appear that my reasoning in developing my operational measure for the concept of intelligence (headsize) is flawed somehow. *My tapemeasure measurements, though reliable, lack validity*, as compared to other generally accepted indicators of intelligent behavior.

V. Statistics, Science and the Scientific Method

Statistics, like mathematics, is a "handmaiden" to the sciences. That is not to say that there is no relevance of statistics to most people's everyday lives. Indeed, we are inundated with statistical facts and conclusions at work, in the daily paper, and so forth. It is rather to say that most *formal* statistical work occurs in the context of scientific investigation, surveys, or research of some sort.

I hasten to add that in this discussion, as well as other portions of this book, I consider "science" to be defined very broadly. In my view, almost any activity that involves the systematic observation of phenomena, particularly if that observation should result in numerical data, is fundamentally scientific in nature. Thus, I do not restrict the term science to the traditional physical sciences, or even to social sciences. What appears to me to be of more importance is the *way* in which the data are developed. If what we call the "scientific method," more or less as described below, was used, then I consider it scientific, regardless of the particular context or subject matter.

Let us now digress slightly to look at the broader context of science, for this is where the conceptualization of the statistical problem is most likely to be rooted; science and research of one sort or another is the wellspring of the data for which statistical methods are the simplifying process.

A. Some Scientific Contexts for Statistics

I will make no attempt to "cover" science, and indeed, do not intend to deal with the specifics of any particular scientific discipline. Rather, I am concerned with establishing the context for many statistical applications in terms of the role played, the essential nature of science and the scientific method, experimentation, and how all of this fits together.

If there is one thing that seems clear, it is that science in all of its forms occupies a very broad role in our daily lives these days. Some 200 years ago, America was about 95% rural and agrarian. At that time, people understood most of the things with which they dealt, since, though life may have been tough, most of it was relatively simple.

A hundred years later, things were not quite so simple, though the majority of the populace still worked the land. Manufacturing was becoming big, the telephone and telegraph were on hand, and electricity was becoming common.

The ever-accelerating march of science and technology over the last hundred years, however, has taken all but about 4% of us off of the land. It has instead submerged us in aeronautics, astronautics, nuclear physics, genetic engineering, TV commercials, VCRs, computer technology, and all kinds of magical-seeming electronic devices. It has become harder and harder to achieve and maintain a clear understanding of how the factors having a vital impact on our lives work. And the pace of scientific development is still accelerating.

Because it is clear that the roles of science, technology, and data in our everyday lives are large and looming larger, because science is a prodigious consumer and producer of data, and because statistics is a discipline designed to handle large quantities of data, a few words about the scientific process and the role of statistics in it seem desirable.

Let me simply cite a few instances in which scientific/research-oriented endeavors produce statistical data: Studies of national health statistics, vital statistics (marriages, divorces, births, deaths, etc.); agricultural production; manufacturing quality control studies; election, political, and other polls; analysis of telemetry data; highway safety statistics; epidemiology studies, entomology studies, etymology studies; studies of animal learning; traffic pattern studies; school attendance studies; economic studies; census studies and reports; ... In short, wherever scientific and/or other human activities result in the production of large quantities of data, there are applications of statistics to assist in extracting the information contained in the data.

B. The Scientific Method

Generally speaking, scientific studies follow a series of steps called the "scientific method" in their attempts to gather or produce data to illuminate some problem of interest. The scientific method goes back at least to the Golden Age of Greece. While it has assumed many and varied forms over the years, it is essentially comprised of a number of steps that combine to maximize the likelihood of enhanced scientific knowledge. Honorable men differ in their statement of these stages of the scientific method, but, generally, the process goes something like this:

1. Problem conceptualization

A problem is identified and conceptualized. The problem may arise in a variety of ways, such as the informal result of practical observation (The chickens are dying off this summer, I wonder why?); or, formally, as a hypothesis derived from a theory (having established his theory of relativity, Einstein hypothesized that light from a distant star would be deflected by the gravity field of a massive object such as our sun). However, in either case, the important aspects of the problem must be conceptualized and explicitly stated in a way that makes systematic investigation possible.

Science is data-oriented. It proceeds on the basis of observations. Unless the thing to be studied can be observed in a meaningful way, it cannot be the subject of scientific investigation. Thus, the scientific method contrives to treat phenomena or problems of interest so as to produce observational data. Thus, part of the problem definition concerns agreement on the specific aspects and conditions of observation.

One problem with this approach is that some things are not easily observable. For example, human personality traits, such as intelligence, are not directly observable (what does intelligence look like?). It is therefore sometimes necessary to agree upon some set of things that *can* be observed as acceptable indicators or proxies for the underlying, unobservable trait.

As we have noted earlier, this process is called the development of an "operational definition." Then the problem is redefined in terms of this operational definition. Simply put, this requires developing a logically defensible linkage between the trait (or the problem to be studied) and those observable characteristics about which we will collect the data.

For intelligence, psychologists frequently use ability to manipulate the verbal and numerical symbol systems as the proxies for intelligence in the operational definition. Their rationale is simply that these abilities can be demonstrated to set human functioning above that of all other creatures, and seem to relate well to human life successes that we think of globally as resulting from "intelligent" behavior. This is why, although psychologists may disagree as to exactly what the underlying trait of intelligence really is, most intelligence tests are comprised largely of symbolic material such as word problems, definitions, analogies, number series, and the like.

2. Data collection

Once the problem has been defined operationally, the scientific method proceeds to the systematic collection of evidence through the application of some form of observation or measurement, producing a data record for analysis. Space does not permit cataloging the variety of measurement formats. They might include questionnaires, tests, telemetry records, temperature records, ionization data, incidence of infections, birthrates, etc.

Statistical issues may be a consideration at this stage; things such as sampling, response rates, and planned treatment of the data may influence the data collection procedures. However, in any case, the data should be systematic and recorded if possible, as opposed to anecdotal, and the methodology of the process of collection must be made explicit. Care is taken to see to it that extraneous data and prejudical conditions are controlled or avoided.

3. Data analysis

This stage requires the application of logically acceptable procedures to the problem of deriving information and conclusions from the data set. Again the procedures of statistics commonly come into play at this point in many studies, regardless of the subject matter discipline. Care must be taken to exclude preconceptions (the tendency of all persons from the man in the street to the aspiring graduate student to "find what they're looking for," whether it was there or not, is notorious).

Care must also be taken to guard against the intrusion of personal feelings into the objectivity of the findings. One may not like the result, but if the data say that something is true, the scientist must accept it—like it or not!

4. Dissemination and replication

Reporting and dissemination of the results are essential. Science cannot proceed without them because they alone permit "independent verification and replication." It is a canon of scientific methodology that any competent scientist should be able to replicate a colleague's study and thus verify his/her results. It is this step of replication and verification of disseminated results that protects science from the idiosyncracies, mistakes, and misperceptions of a single worker's efforts. This is why the scientific/academic/research community places so much emphasis on publication.

Thus, the report must contain not only the findings of the study, but also a complete description of the methodology and the conditions under which the study took place—and any other information necessary for independent replication. As noted above, replication is the aspect of scientific methodology that compensates for mistakes and lack of objectivity in the original work. Without it, human foibles being what they are, the body of scientific knowledge would contain many more false leads and misconceptions.

5. Knowledge integration

In addition to reporting the results of the study, the scientist has the obligation of attempting to integrate his or her findings into the larger body of knowledge pertaining to the problem studied. Beyond the immediate outcomes of the study, the findings may fill in some long missing piece of a larger puzzle. They may permit progress toward the next two stages of knowledge development, theory building and scientific explanation.

Theory building occurs when the scientist is able to link together sufficient knowledge, usually from a variety of sources, to be able to construct predictions about related problems. Then new studies can confirm or refute these predictions, and the theory can be further elaborated.

Scientific explanation occurs when theory is sufficiently developed and tested that it begins to assume the character of a Law, and is capable of specifying how the phenomenon works, and under what conditions. Ultimately, of course the objective is to be able to delineate the complete set of causal relationships that trigger the occurrence of the phenomenon and control the outcomes.

C. Experimental Controls and Causation

It may be postulated from the above discussion that the ultimate goal of science is to explain the world in terms of what causes what. In the paragraph on data collection above I glossed over the details of method and procedure in favor of the larger view. Now, however, as an example of one type of scientific approach (a most important one), let us consider the use of experiments to produce data, and the design of controls intended to let us impute weight to the outcome.

Experimental methods are not applicable to all scientific problems. As the name implies, experiments are events particularly designed for the purpose of investigating some hypothesis or problem, and since they are so designed, they must be of manageable scope. (The major alternative is "in vivo" work, or taking it as it comes. Astronomical studies are an example of the latter type, since no one has figured out how to subject astronomical phenomena to a design and control process.)

Laboratory studies are frequently of the experimental type because they can usually be closely controlled. Fundamentally, the steps involved in experimentation are those of the scientific method. The essential addition, however, is that the design involves the ability to control what goes on and to manipulate one or more of the factors bearing on the outcomes.

The most basic experimental design says that if you can identify every aspect of a situation that makes any difference in what happens, and if you can hold all of them but one constant (unchanging) while changing the one in a known way, then whatever happens must have resulted from (was caused by) the change you made.

This seems entirely reasonable! But, if we know so little that we must run studies on this thing, then it is unlikely that we will know enough to say that we have identified all possible contributors to the outcome. Medical research, for example, is always concerned about the impact of unknown factors on an experiment—a virus as yet unidentified, or unforeseen interactions between medications, or unknown genetic predispositions in the subjects, and so forth.

And, what about holding everything constant in a changing world? If it is an experiment on human subjects, factors such as age and intelligence and ethnicity and knowledge of calculus can be controlled, but how do you allow for mood changes due to the weather on the day of the experiment, or loss of sleep the night before?

Even factors that appear to be constant can change without the knowledge of the experimenter. Juries are sequestered to shield them from knowledge input during the course of a trial which might change their reactions to the case; but, even the slightest bit of seemingly innocuous information might cause some juror to respond differently and prejudicially to some aspect of the case without anyone being the wiser. These problems exist for all kinds of experiments from physics to social psychology to education.

Then there is the problem of being able to change, in a known manner, just one of the constellation of applicable factors. In an experiment that I did one time on tactile perception of colors, red and white poker chips were used. The focus of the experiment was to determine the subject's ability to sort the poker chips into red and white piles by touch alone. Unknown until later was the fact that the red dye used in the otherwise similar poker chips had perceptible tactile characteristics of its own. Thus, instead of manipulating one factor (color) in a known way, additional changes were introduced (the tactile characteristics of the dye) which ruined the experimental design.

Just the sheer act of measuring something sensitive can influence the outcome and ruin the experiment. Scientists have long known this effect and take extraordinary precautions to avoid it. A simple example comes from an attempt to measure the amount of heat in a chemical reaction. using a lab thermometer. Inserting the thermometer into the mix can result in the thermometer absorbing some of the heat of the reaction, thereby causing an underestimate of the total heat produced.

Subjects in pyschological experiments frequently become aware of the point of the experiment and, unknowingly, begin to react in the way the experimenter expected. Humans are bad news in this respect. Just the fact that they have been tapped to participate in an experiment sometimes results in improved performance, obscuring the impact of the experimental factors under study (this is known as the "Hawthorne Effect," after a famous study of worker productivity carried out at the Western Electric plant in Hawthorne, NJ, in the 1930s; see Roethlisberger in the reference list).

Finally, there is the problem of being able to detect the outcome in the first place. Physicists have reached such levels of delicacy and work with such minute and rare events that many times theoretical developments await the invention of a detection device sufficiently sensitive to record the outcome of the experiment.

Psychologists have a problem of a different nature. In their case, the problem is frequently to be able to sort out the desired happening from among all of the other ongoing aspects of human behavior. And, in the last analysis, discerning the happening and measuring it is to no avail if there is a flaw in the logic linking the thing observed (the operational definition) with the conceptualization of the underlying phenomenon of interest.

D. Some Control Strategies

In view of all of these problems bearing on the use of the basic experimental design, scientists/researchers adopt a number of devices in their attempts to achieve their ends. In general, they all have to do with the subject of "control." That is, they are ways intended to control the errors that would be introduced by unknown factors and by biases of various sorts, all in the interests of getting a more accurate view of causal relationships. A brief overview of some of these control strategies follows

1. Matched pairs

The tightest design, theoretically, is one where two groups of items or persons or whatever are selected at random from the population of interest in such a way that each item in one group has a counterpart in the other that is alike in all significant ways. Then the experiment is done (one factor is changed) to the members of one group and not to the other. At the end of the experiment, measurements for the two groups are compared, and if there is any difference, it is presumed to be attributable to (caused by) the experimental procedure (the one changed factor).

Of course, the practical problem is that exact matches are hard to achieve, and even harder as the number of factors on which one must match increases. Let's go back to raising fruit trees for example. Suppose we divided the fruit trees in an orchard into two groups. Each tree in one group would have a matching tree in the other group. Matching factors might be such things as variety, size and age of tree, history of disease, soil analysis and Ph, frequency and timing of pruning, etc. The experimental factor might be a standard amount of fertilizer administered only to one group. Yield would then be examined for the matched pairs under the hypothesis that the fertilized member of each pair should exhibit higher yield.

2. Matched groups

An approximation of the matched pairs method of controlling the factors in an experiment is the use of matched groups. This method is based on the same logic, but recognizes that individual pairings may be so difficult or impossible to achieve that a match will have to be done on group characteristics rather than individual ones. Probably the same factors are used, but instead of requiring a tree by tree match, one is satisfied when the match is accomplished according to the average status of each group with respect to each factor. For example, we would require that the *average* age of the trees in each group be the same, that the *average* soil analysis be the same for each group, and so on.

Other things being equal, this is a weaker design, because within the averages on any particular factor there could be considerable differences between the experimental individuals and the control individuals which don't enter into the results because they are lost in the averaging process. Thus the attribution of causality to the experimental factor is on less sound ground.

3. Random groups

The process of matching, even on a group basis, is sufficiently difficult and costly that it is often not possible. An alternative is simply to pick two groups independently and at random, trusting to the random selection process, or chance, to result in two equivalent groups. One is then subjected to the experimental treatment while the other is not, and comparisons are made on the outcome.

This process could not be used to compare apple trees in two different states such as Washington and Virginia, because the Virginia and Washington trees already exist in natural groups. If, however, one were to try the fertilizer experiment in Virginia alone, one might identify a large grove of trees for study, and then randomly assign each tree in the grove to one or the other of the experimental and control groups.

Because this random assignment involves only indirect control of those factors that might be significant in influencing the outcome, it is even less strong than the group matching. *If random process fails by chance to balance an important factor between the two groups, then differences ostensibly the result of the treatment will really arise elsewhere.*

Worse, it is not possible to be sure just what really happened. There are statistical treatments that can be applied with respect to a few factors to adjust for important differences in the control factors between the groups, but these treatments are complex and limited in scope.

4. Self-comparisons

Another strategy is to select a group at random and use it as its own control. This is the before and after plan. The items are measured before the expermental treatment is administered, and then again afterwards. This plan controls for all of the factors that might be associated with the comparison of different individuals, but leaves uncontrolled any factors associated with the passage of time. These might include personal growth or learning between first and second measurement, as well as the impact of any uncontrollable or measured event that might have taken place in the interim.

To go back to our apple trees, suppose the Virginia grower had decided to run the fertilizer experiment on a single orchard. Noting the yield for the season with fertilizer is higher than it was for the previous season without the fertilizer, he decides that the fertilizer is a good idea. Unless he is careful to allow for the effects of differential rainfall during the two seasons, a later first frost, diseases and pests, and other such factors, the exact impact of which is hard to estimate, he may well come to a false conclusion.

5. Summary

The above overview presents the most common strategies used by scientists in their efforts to tease out causality in the experimental setting. It should be noted that none is totally satisfactory for all applications, and that all have significant problems from the point of view of practical applications. This perhaps partly explains why the progress of science toward scientific explanation tends to be fitful and erratic, and frequently dependent upon the development of new statistics and techniques for designing experiments and collecting and analyzing data.

VI. Statistics Is for Everyone

It should be clear by now that it is my position that the science, research, and statistics used to advance human knowledge are of the greatest importance, at least indirectly, to every person. Not only that, but we all deal increasingly with scientific and numerical data in our everyday lives, both on the job and at home, in the form of information, education, tools, processes, and procedures.

It becomes increasingly necessary for us to act in some ways in response to such things—voting for instance, with respect to scientific ideas and proposals. Should we support the national space program, the development of new defense initiatives, or genetic engineering approaches to controlling crop disease and fixing human defects? These decisions become a little easier if one has some feeling for their scientific/numerical underpinnings. If one can read a report and have some confidence that the principles of the scientific method apply, rather than bias and inobjectivity, or worse, raw prejudice and greed, this can be comforting and even advantageous. Just a familiarity with some of the basic statistical principles can be of help.

Then, too, each of us is a statistician in his or her own way. When you cross the street, you make an estimate, based on the data of experience, that there is an acceptable probability of reaching the other side intact. You might be hard put to state it as a decimal fraction, but it is there. We all use the main ideas of statistics at the verbal level. Ideas such as "the average man"; "people vary a lot"; "those things are related"; and approximations, and guesses about how accurate they may be, and so forth. In the next few chapters we will try to extend these concepts a bit, bringing a little more substance to them without wallowing in technical detail, so that they may perhaps become a little more useful to you in a broader range of situations.

VII. Summary

In this Chapter, I have tried to establish the fact that things statistical in nature are found in many arenas of endeavor and have impacts on the personal lives of all of us. I have tried to describe and delimit the nature of statistical activities, and in general establish a context for statistics as a foundation for the more specialized discussions found in later chapters. Some of the major points made were:

- "Statistics" may refer to a discipline, or to statistical data, or to specific indices resulting from the application of statistical procedures.
- Statistical procedures apply only to large masses of data, usually in the form of numbers (the results of counts or measurements).
- The functional numbers used in measurement and statistics tend to be of three general types: nominal, ordinal, or cardinal.
- Statistical procedures are generally of three types: descriptive, inferential, or relational.
- Overarching statistical concepts that may be found in connection with many different statistical activities and procedures include average, approximation, probability, randomness, and individual differences.
- The nature of measurement was discussed, and it two most important characteristics, reliability and validity, were described.
- Formal applications of statistics are generally found in connection with research and sciences of various types, but statistical ideas are everywhere in daily life.
- The basics of the scientific method were discussed, since statistical applications occur so frequently in scientific endeavors.
- A major scientific method, the experiment, was discussed, and several procedures used in experiments to "control" various sources of error and bias were summarized.
- It was noted that we are all statisticians in our personal lives at some, perhaps rather unsophisticated, level.

VIII. References

Baskin, John T. *Probability: A Noncalculus Introduction*. Totowa, NJ: Roman and Littlefield, 1986.

Edwards, Allen L. *Experimental Design in Psychological Research*. New York: Harper and Row, 1985.

Fabian, Vaclav and Hannan, James. *Introduction to Probability and Mathematical Statistics*. New York: Wiley, 1985.

Flanagan, John C. *The Aviation Psychology Program in the Army Air Force*. Washington: U.S. Government Printing Office, 1948.

Herdon, Peter, *Statistics: A Component of the Research Process*. Norwood, NJ: Ablex Publication Corporation, 1991.

Lastrucci, Carlo L. *The Scientific Approach: Basic Principles of the Scientific Method*. Cambridge, MA: Schenkman Publishing Co., 1967.

Linn, Robert L. (Ed.). *Educational Measurement, 3rd Ed*. Washington: American Council on Education, 1988.

Keedy, Mervin L. and Bittinger, Marvin L. *Introductory Algebra, 2nd Ed*. Reading, MA: Addison-Wesley, 1986.

Roethlisberger, F. J. and Dickson, W. J. *Management and the Worker*. Cambridge, MA: Harvard University Press, 1939.

Ross, Sheldon M. *Introduction to Probability Models, 4th Ed*. Boston: Academic Press, 1989.

Schmitt, Neal, Borman, W. C., et al. *Personnel Selection in Organizations*. San Francisco: Jossey-Bass, 1993.

Spearman, Charles, *The Abilities of Man*. New York: MacMillan, 1927.

Stigler, Stephen M. *The History of Statistics*. Cambridge, MA: Belknap Press, 1986.

Williams, Bill. *A Sampler on Sampling*. New York: Wiley, 1978.

II

Basic Statistics

Now that we have reviewed numbers and their basic manipulation, and have looked at what statistics is and does and where the numbers come from, it is time to begin the treatment of basic statistical topics: arrays and distributions, averages, and variability. Chapters 3, 4, and 5 take up these topics, respectively.

These three topics comprise the most fundamental of those generally covered in the traditional statistics texts. Here, however, we will emphasize explanations rather than the computation and display of sample problems.

These three topics comprise the most simple, straightforward statistical treatments that can be applied when one is faced with a large quantity of data. They are generally attempts to answer three basic questions, respectively: What do the data look like? What is a "typical" value? How much do the individual observations spread out from the middle?

Before going on, I would like to clear up a possible source of confusion. The question is how to refer to the values that make up a set of numbers upon which we would like to perform some statistical operations. Obviously, they might take different names according to the problem and the source of the numbers.

In the next few chapters, I will use several designations, but most likely I will say "score" or "scores." The term is used in a general sense, and not restricted to sports, games, or educational achievement tests. Sometimes I will say "value," sometimes I will be able to use the name of a particular measure, but most often I will use these interchangeably with the general term, "score."

3

Arrays and Distributions

I. Overview

In this chapter we begin to consider things more statistical. What if you should suddenly be confronted with a large mass of data, numbers, and a need to understand the information hidden therein? What would you do?

In this chapter we define certain terms, and then confront this problem. Techniques for teasing out some information are presented, including drawing a picture of what is happening in the data. I also show you some of the "standard" or "typical" pictures that result from graphing various types of data, together with some of their meanings.

One of the cardinal rules in dealing with data can be summed up: "Get your hands dirty in the data." The importance of this admonition grows daily. More and more numerical/statistical processing is not done by the researcher, the consumer, or by the person who compiled the data. Rather, it is done by second and third parties, on machines that may not even be physically present, using immense and complex programs designed by yet other persons.

Such conditions tend to promote in the user a remoteness from the reality that the data are supposed to represent. They contribute to a sort of dissociation with the quality control which any data user should impose on his or her numbers and the interpretations thereof. Certainly, these aids have made possible analyses that could never have been done in an earlier time, and at the very least have removed much of the burden of processing the findings from the researcher and data user. However, you must be very careful to stay in close touch with your data. Otherwise, you may inadvertently fall prey to errors and misinterpretations arising from a lack of familiarity with the real data and the actual conditions under which they were collected and processed.

One of the most important ways in which the data user "gets his hands dirty" is to examine how the individual measurements relate to, or "distribute" themselves along, the dimension or scale being used. This is done by arranging the data in various ways and by drawing pictures of the relationships in the data. Through "distributional" considerations and graphing (pictures) the researcher can acquire a feeling for his or her data which often enables him or her to spot meanings (or errors) that might otherwise go unnoticed in the vast mechanical morass of large-scale data handling and processing.

It is the purpose of this chapter to examine some of the various ways in which information in the data may be depicted, and the sorts of information that may be derived from these efforts.

II. Arrays and Distributions

We will start with some terms and their definitions.

A. Some Terms of Interest

Although some of these terms were mentioned in Part I, it is best to summarize them here, for they play an important part in the process of "getting a picture" of what is really happening in your pile of numbers.

1. Population or universe

No matter what one is studying, there is some set of items, objects, or individuals in which one is interested. This set is called the universe, or study population. (Some authors distinguish between population and universe, whereas others use the terms interchangeably; it will not be necessary for us to distinguish.) The population is always carefully defined by specifying as completely as possible those *critical attributes* that distinguish its members from any and all other entities. This step is essential—unless it is carefully and accurately done, it will be impossible to designate unambiguously *from whom* the data are to be collected or *to whom* the results apply.

For example, suppose you wanted to study the school children in Bonhomme County. To define your population, you would have to make decisions such as: Do you include all children on the school rolls, or just those in attendance on a specific day? Do you include homebound children? Home instruction children? Those living outside the County but attending school within? Those living within, but attending school outside? Nursery schoolers? Kindergarteners? Daycare? And so forth.

Once the population has been defined, then only those who meet the definition are eligible to be included in the data collection, and, conversely, whatever conclusions result apply only to that particular population (those who meet the definition). It is not permissible, for example, to study the children in Bonhomme County, however defined, and apply the results to the children of adjacent Harmony County.

2. Sample

Sometimes, for theoretical, practical, or efficiency reasons, it is desirable to study (collect data from) less than the entire population. In such cases a subset of the population called a sample is selected (using means and strategies that are considered in a later chapter). Although data are then collected only from or about the sample, conclusions are drawn (generalized) to the larger population as well.

What is the essential nature of a sample? In a word, a sample should be "representative." This means that, effectively, a sample should be a small-scale replica of the population from which it is selected, in all respects that might affect the study conclusions.

3. Variable

A characteristic, quality, or trait with respect to which the members of the population differ (vary) among themselves is called a variable. In statistics, "x" and "y" are frequently used as literal numbers to represent particular variables (though any literal number might be used). As usual, we employ the literal number so that we may talk about the variable in the abstract.

Sometimes the term "dimension" is used essentially synonymously with variable. It is usually continuous in nature, represents some underlying construct, and is the subject of a measurement procedure (using some kind of scale) resulting in data. It may exist in nature, or it may be a statistical invention constructed to aid in the measurement or interpretation of the construct.

When data are gathered on the "x" variable, this means that each member of the population from whom information about "x" was collected has a specific number as his "x-value." The set of these values may be collectively referred to as the "xs," and if it becomes necessary to refer to them separately, for individual members of the population, frequently subscripts are attached to designate which individual observation is being discussed; for example, $x_1, x_2, x_3, x_4, \ldots$

4. Constant

A constant is a trait or quality for which the members of the population (all of them) do NOT differ or vary. In statistics, the letter "k" is often used to represent the numerical value of a particular constant (also, sometimes, "*a*", "*b*", and "*c*"). If there are several different constants to be considered, different letters may be used; but if different values of the same constant are intended, they are usually distinguished by the use of subscripts, for example, k_1, k_2, k_3, etc.

5. Distribution

When a set of measures, which vary among the members of the study population, has been collected, we know that they consist of differing values. It is then legitimate to examine the phenomenon of their difference. When we have, in some fashion, laid the individual members of the set of measures out for examination, we have created a "distribution." It is the distribution, revealing the extent and nature of the differences in the set, which provides the "picture" of the data to which I referred earlier.

B. A Basket of Numbers

Suppose that we have identified a population in which we are interested, developed a construct, and used it to guide the collection of data about this characteristic (call it "*x*") from a large number of the members of our population. Now we are faced with the problem of extracting something meaningful from this mass of measurements. In other words, if we have a huge basketful of *x*s, where *x* is a variable, how would we go about making sense of it? Answering this question will preoccupy us over the next several chapters. The first step is to examine distributional considerations.

1. Distributional considerations

There are several characteristics of distributions that might be examined.
   a. Number. You would begin to make sense out of your basketful of numbers by trying to answer some basic questions about them. The first might be, "How many do I have?" If you were to count the numbers in the basket, you could answer this question, and you would have your first piece of statistical information about the set.
   We could call this "number," and for convenience might abbreviate it "*N*."
   Thus, if someone should ask, "What is your N?", we could say $N = 8178$, or whatever.

But, this is so simple. Does it help? Does it mean anything? Probably not much, by itself, but taken with other bits and pieces it could be exceedingly important. Suppose you have determined by other means that there are 10,000 items in this particular population. If N is 8178, it means that you have measured by far the majority of them, and thus, that any conclusions you may draw are far more likely to apply to the whole group than if you had measured, say, only 85 of them. Not unimportantly, N also gives you a basis for estimating the processing cost of dealing with your basket of numbers in other ways.

b. Array. The next question to be addressed might be, "How big are the numbers?" You might sift through the entire mass of numbers noting the big ones and the small ones, perhaps looking for the biggest and the smallest. Of course, if you found them you may later be asked for the next biggest and/or the next smallest, and you would have to start all over again. After spending a lot of time on such questions, you would probably opt for simply arranging the set in either ascending or descending order (which would enable you to answer all such questions). Such an ordered arrangement is termed an "array."

Again, this is a simple procedure—rather tedious with 8178 numbers, but it can be done. What have we learned? This time we have learned much more. Not only have we identified the largest and smallest of our numbers, but we can also identify those portions of the dimension measured where there are lots of numbers (many members of the group had these particular values), and those where there are few or none.

In other words, our array permits us to determine the relationship between the scale values on the dimension and the frequency with which those values occur in our basketful of numbers. By constructing the array we can see how the data are "distributed" with respect to the underlying construct that gave rise to the data.

Of course, though conceptually simple, from a practical standpoint, arraying 8178 observations is a monumental task. Consider this: if you could get 1000 of these observations on a single sheet of paper it would still take you more than eight sheets of paper to complete your array. In addition, although all of the information about how the observations were distributed along the dimension would be there, it would be enormously difficult to comprehend, simply because of the sheer volume of numbers to be examined. Surely, there must be a better way and, of course, there is.

c. The frequency distribution. To simplify the process of extracting information from our huge array of numbers, let us consider a special kind of array (remember, an array is simply an ordered list, from largest to smallest, or vice versa).

Note that in measuring any usual dimension, across a substantial sample or population, we may expect to find many duplicate values in the data, that is, scale values that were assigned to more than one member of the measured group. To take an extreme example, if your scale were "Male or Female," you would expect to find approximately half of the measured group attaining one scale value (Male), and half attaining the other (Female).

A 3-point scale of "smarts" would probably get about a quarter of your group in the "Not so," about half in the "So-so," and about a quarter in the "Very" categories.

To go back to our problem, if, instead of arraying all 8178 observations, we arrayed only the set of *possible and different* scale values, we could then distribute each one of our 8178 observations by assigning it to one of the possible and different values in the array. Then we could tally for each one of these the number of times it occurred.

In effect, then, we would still have an array (of possible and different values on the scale). But now we would have grouped the observations by tallying each one against one of the possible, but different values. In doing this, we would still be accounting for all 8178 observations. Such an array is called a "frequency distribution"; it is an array of all of the possible different values, with a count of the number of occurrences (or frequency of occurrence) for each different value in the array.

To take a hypothetical (made up) example, suppose our measurement dimension had been "Number of Brothers and/or Sisters Living," and we had used a 10-point scale as shown in Table 3-1. If we gathered this piece

Table 3-1. Number of Brothers and/or Sisters Living

Number Living	Frequency (f)
9 or More	83
8	57
7	82
6	170
5	361
4	542
3	1066
2	2039
1	2656
0	1122
Total	8178

of information for each of the 8178 members of our sample, and arrayed this information according to our 10-point scale, we might get a frequency distribution such as that shown in the table.

Notice how much more compact and informative this arrangement is than it would be to array all 8178 numbers individually. Indeed, this frequency distribution makes clear several characteristics of the data set in a single glance. We can see instantly that higher numbers of living brothers and/or sisters is an increasingly rare event; that a substantial number of our group have no living brothers or sisters; and that a small but noticeable number of our group have very large families, with 9 or more living brothers and/or sisters.

It would have been all but impossible to glean this much information from the original array—although, obviously, it was all there. Thus, the frequency distribution usually has the virtue of compacting the data to a more manageable scope.

d. The grouped frequency distribution. You will have noticed that, as we have approached the problem of making sense out of our basket of numbers, both arrays and frequency distributions have had the effect of reducing the enormity of the problem of comprehending and digesting the information contained in the data. They have accomplished this by introducing order and by reducing the sheer number of bits of information to be absorbed at one time.

We are now going to take this process one step further. Instead of distributing the frequency of occurrence of the individual scale values, we will first group those scale values into a set of (usually) 10–20 clumps of values. Why?

To explain this, it is necessary that you understand that there are quite frequently situations in which even the regular frequency distribution as described above becomes unwieldy and too difficult to handle. As an example, suppose you work for a school district and you want to deal with students' SAT (Scholastic Aptitude Test) scores for the district.

The range of the SAT distribution is from 200 to 800 points on each of the two main tests, Verbal and Mathematics. Over a population the size of a school district, most of those possible values would be obtained by some student(s), and your frequency distribution might have to have 500–600 scale values in it. Again, this is far too much for ease of handling and comprehension.

But, suppose instead of counting the frequencies for each individual value, we first lump some of the values together and then count? For example, we might decide that 50-point intervals would be sufficient discrimination among the students to serve our purposes. This might give us a (grouped frequency) distribution such as the following:

Scale Intervals	Frequency
(SAT Verbal Score)	(f)
750–800	(tally)
700–750	"
650–700	"
600–650	"
550–600	"
500–550	"
450–500	"
400–450	"
350–400	"
300–350	"
250–300	"
200–250	(tally)
TOTAL	(sum)

If we use a grouped frequency distribution, what have we gained and what have we lost? It is clear that it is a large advantage to be able to see all of the distribution of frequencies on one page instead of the 12–15 pages we would have to use if we had opted for a simple frequency distribution. Indeed, such a compacting of the data sometimes reveals major trends in the distribution that would be somewhat obscured by minor fluctuations from step to step in the scale. On the negative side, should such narrow fluctuations represent something important in the data, we have lost it forever because once the data are grouped, there is no way to dissemble them into their original singularities.

However, usually, the minor fluctuations result from biases, random factors, and errors of various kinds, and by and large we are better off without them. Thus, grouped frequency distributions tend to convey as much real information as do simple frequency distributions.

e. Rules of thumb for grouped frequency distributions. If you want to make up some grouped frequency distributions of your own, there are a few rules of thumb to be considered:

(1) How many groups (also called "intervals" or "score intervals") should there be? As suggested above, it would be convenient to get the distribution into a form that can be shown on a single page; also, one should not have a great number of small intervals, as these tend to

overemphasize the minor fluctuations in the data just as the simple frequency distribution would.

But, on the other hand, one should not lump everything into just a handful of intervals, as this would tend to mask real differences in the data. So, how does one know? There is no substitute for a real and deep understanding of the phenomenon being studied to provide the answer to how many; but, if you don't know, and you have to have a rule of thumb, somewhere between 10 and 20 is suggested.

(2) If the variable is continuous, the set of intervals should be continuous; that is, the intervals should cover the score scale from the lowest obtained value through the largest, with no gaps (even if some intervals have zero frequencies). This is because the interval set is intended to represent the score scale and thus should cover the same territory.

(3) It is much to be preferred that the intervals all be of the same size. If you decided to lump together 5 score values into an interval, use 5 as your interval size at all parts of the scale. This is because calculations are sometimes based upon grouped frequency distributions, and these calculations are greatly complicated by differential interval sizes; also, if the distribution is plotted, the difference in interval sizes along the scale can greatly distort the picture being presented.

(4) If possible, stay away from "open-ended" intervals (such as "240 and up"), because, for either computation or plotting, it is impossible to fix the midpoint of such an interval. (As we shall see presently, midpoints figure prominently in drawing a picture based upon a grouped frequency distribution.)

(5) The set of intervals should be "mutually exclusive and mutually exhaustive," that is, when you go to tally the scores into the interval system there should be a place to tally *each* score, but *only one* place. If you look back at the distribution of SAT scores presented earlier, you will see that I have violated this rule, and that possible difficulties could arise because of it. For example, take the intervals 250–300 and 300–350; in which one do you tally a student whose score is exactly 300?

In order to avoid this kind of quandary, we assume that intervals begin and end at places where no real scores will fall. We usually use *halfway-points* between the upper end of one interval and the lower end of the next higher interval. In this case, then, the first interval would begin at 250.5 and end at 300.5, and the second would begin at 300.5 and end at 350.5.

Thus the score of 300 belongs unambiguously in the first interval, and not in the second! Also, even though the value of 300.5 occurs in both intervals, there is no problem, *since half-score-points do not exist among the real scores obtained by the students.*

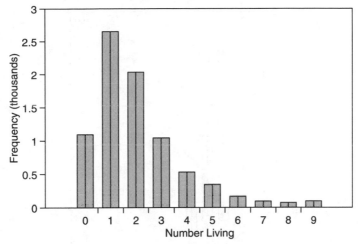

Figure 3.1. Brothers and/or Sisters Living

2. Graphing the frequency distribution

The information in frequency distribution data may be even more easy to see if we convert the distribution into a picture. In order to do this, we set up a two-way plot, with the scale values across the bottom, arranged in interval fashion (equi-step), and the frequencies (counts of the individual values) arranged in equi-step fashion according to the vertical dimension, so that by going to a particular scale value, the appropriate frequency is given by reading the height along the vertical scale of frequencies.

Our simple frequency distribution table for number of brothers and/or sisters living is shown in graphic form in Figure 3-1.

This graph, or picture, of the distribution is called a "bar," or "vertical bar," graph. The shaded parts, or bars, are used to indicate the number of frequencies (the counts) for each of the scale values shown along the bottom (or horizontal axis). The height of each bar is proportional to the number of frequencies counted up for that particular scale value.

With the large numbers of counts in the distribution of our example, it is of course impractical to make the vertical scale large enough to show a single count—but this is acceptable, as any differences dependent upon just a few counts would undoubtedly suggest spurious accuracy. Of course, if we wished to use up more space, we could use a larger scale on the vertical axis; here I have used about 1050 counts to the inch, (at least that is what it was before printing) but if we wish to stay on a single page we couldn't stretch the scale to more than about 350 counts to the inch.

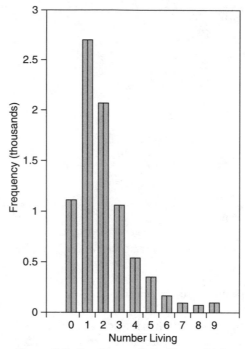

Figure 3.2. Brothers and/or Sisters Living

It would be possible to change the picture presented by this graph considerably *without* changing either the form of the graph or the data (distribution shown in the table). Expanding or contracting either scale relative to the other would change the impression created by the picture.

For example, as shown in Figure 3-2, the vertical scale has been stretched while the horizontal scale has been reduced. This maneuver overemphasizes the frequency differences between adjacent scale values.

On the other hand, as shown in Figure 3-3, I have used just the opposite tactic to deemphasize those differences.

So now, you see one of the many ways it is possible to "lie" with statistics—just by using the scale characteristics of the graph to create differing impressions.

In the foregoing comments I have been discussing the simple frequency distribution presented earlier. Of course, *by plotting the frequencies above the midpoints of the scale intervals*, graphs for grouped frequency distributions can be prepared just as with simple frequency distributions.

Further, the bar charts referred to above could be rotated 90 degrees clockwise to become a "horizontal bar" chart. The same information could

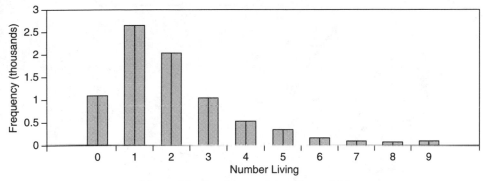

Figure 3.3. Brothers and/or Sisters Living

also be presented in a "line" graph, by placing a dot at the appropriate height over each scale value and connecting the dots with a line. Figure 3-4 shows the original bar chart above in line graph form.

Most of the distributional plots presented in this chapter are either line or bar charts, but we might digress for a moment to discuss two examples of another type of graph that is frequently used in presenting statistical information, "area" charts.

a. Pie charts. "Pie" charts are area charts that are frequently used where there is a logical "whole," and we wish to show the relative sizes of its various pieces. Let's take a corporate budget as an example:

Category	Expenditure
(1) Salaries and Wages	$100,000
(2) Benefits	30,000
(3) Contracts and Services	15,000
(4) Communications	5,000
(5) Overhead	80,000
(6) Fee	20,000
TOTAL:	$295,000

Figure 3-5 illustrates this budget in the form of a "pie" chart. The size of the pieces is derived from taking the 360° of the circle (pie) and apportioning it into "slices" according to the relative sizes of the various budget items. That is, the pie is divided proportionately according to the relationship that each item bears to the whole pie. Pie charts implicitly represent percentage distributions, which by definition add up to a whole (100%).

Although the number of degrees of the circle involved in each piece is linearly proportional to the data represented by that piece, the visual

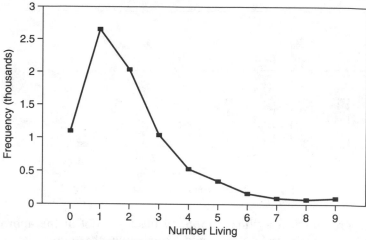

Figure 3.4. Brothers and/or Sisters Living

impression is also affected by the area of the pieces. Because the area is a function of the radius of the circle (πr^2), it is important when viewing several pie charts that the circles be of the same size (radius, area). If they are not, the impression of the size of a slice is colored by the difference in the area of a slice between the two pies. That is, a 12% slice of a 6-inch pie *looks* a lot bigger than a 12% slice of a 3-inch pie.

Figure 3-6 shows a 12% slice of a pie chart along with a 12% slice of a pie chart with twice the radius of the first one. You can see that if these were shown together, the latter would *look* bigger, even though it represents the same proportion of the total as the former does. The impression of difference arises from the larger area.

It is thus important in dealing with all kinds of charts that have area as a characteristic feature to determine which dimension of the figures being

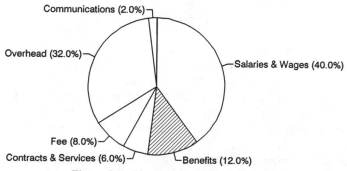

Figure 3.5. Pie Graph of Expenditures

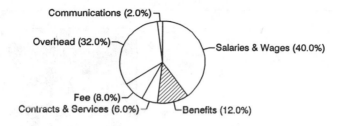

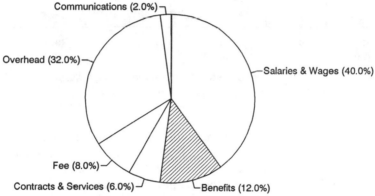

Figure 3.6. Pie Graphs of Expenditures

used is proportional to the data being depicted. Is it the area, the radius of a circle, the height of a figure, or some other measure?

b. Figure charts. The "figure" chart is also an area chart—and even more tricky than the pie chart. Figure charts are frequently really *area* charts; that is, the *impression* that they create is that the various data values are proportional to the area of the figure. However, most of them are constructed so that the proportionality is really between the data and a linear dimension—usually the height of the figure. This introduces a distortion, since the area tends to make the figures look disproportionately larger than they should. Figure 3-7 shows an example of a chart in which the data are proportional to the area of the figures.

This distortion is not so easily controlled as the pie chart, because it is not as simple as standardizing the radius of a circle. One possibility is to vary only a single dimension of the figure. In Figure 3-8 the shaded figures are proportional to the data in height, whereas, for contrast, the open figures are proportional in area. Another possibility would be to use a

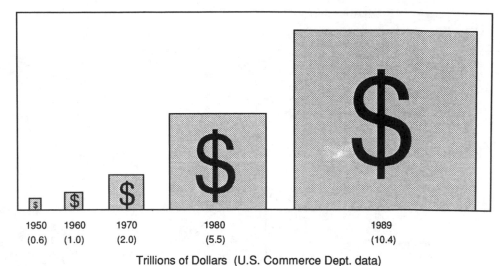

Trillions of Dollars (U.S. Commerce Dept. data)

Figure 3.7. Gross National Product, 1950–1989

single figure and present only fractions of it; this approach is exactly equivalent to using the pie graph. Finally, one could pile up differing numbers of small icons to represent differing amounts; but such an approach is little better than the simpler bar graph we were discussing earlier.

It should be clear now that the apparently simple process of drawing a picture of the data entails some complex issues—always assuming, of course, that you intend to present the information in the most straightforward manner possible! Considerable impact on the impression of the data conveyed arises from the choice of graph type and layout, and from your selection of the number of units per inch (scale) to use in plotting your data on the graph.

c. Some hints about charts/graphs. There are a few hints I can give you about graphs. Most importantly, a graph should be derived naturally from the data to be presented—or at least be compatible with those data. Some more specific suggestions follow.

What are the right scales? This is a difficult question, and does not have a single, unequivocal right answer. Generally, the best answer is that *the data should be presented in such a manner so as not to distort the impression they give*. This will differ from graph to graph as a function of the range of the dimensions, the nature of the data to be presented, and the relationship of the scales to each other, also as a function of the available space and its relationship to the text and the page.

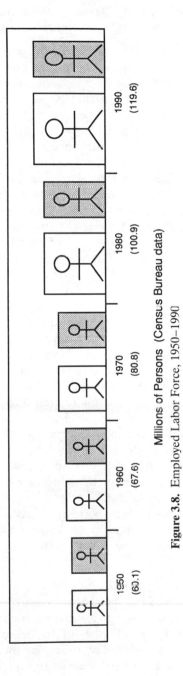

Figure 3.8. Employed Labor Force, 1950–1990

Some rules of thumb can be offered in these matters; however, most apply more to line and bar charts than to area charts:

1. Most graphs use scales for which the zero point of both vertical and horizontal scales is located in the lower lefthand corner of the graph, and which are depicted as continuous lines or a complete sequence of discrete points; that is, there are no gaps permitted in the scales. If this requirement becomes a problem, it may be necessary to rethink the formulation of the data that are to be presented.

It is occasionally permissible to introduce a break in a scale, but this is to be discouraged. If this is done, the possibility of distortion grows large. It is most common in cases where, for some reason, it is desirable to show the zero point on the scale, but all of the data lie far above it. Then, the axis may show a break (done with two wavy lines through the axis), and the same scale as before is resumed where the data points begin.

2. The height of the picture (the plot of the data) is rarely less than 60% of the width, or vice versa. Put another way, the long dimension of the plot, or picture, should rarely be more than 1 2/3 times the short dimension, considering only those parts of the scale for which actual data are being plotted.

3. Show only 1 or 2 scale points beyond the most extreme values for which data actually exist. That is, if you have a 20-point scale, but no one got above 10, you should not show anything beyond 12. To do so would distort the proportions of the picture you are plotting.

4. Good graphs are like good paragraphs—they are generally confined to one idea, or perhaps several closely related thoughts. They should not be overly complex, cluttered, or used to display disparate things.

5. In addition to being confined to limited subject matter, a graph should be limited to a single type of presentation. For example, you shouldn't place a circle chart in the same picture as a bar graph, etc.

6. If a graph will take up more than half a page in the final text, it should be placed on a page by itself.

7. Both the vertical axis and the horizontal axis should be labeled—the viewer of your picture has a right to know what it is that you are presenting. Indeed, graphs should have a title as well, just as a table should.

Further, if there is more than one bit of data being presented (such as a line showing the growth of the male population over the decades of this century, and another line showing this growth for females), the separate bits of information (lines) should either be labeled or coded and a legend presented identifying the codes. For example, the male line might be presented as a series of dashes (----) and the female line as a series of dots

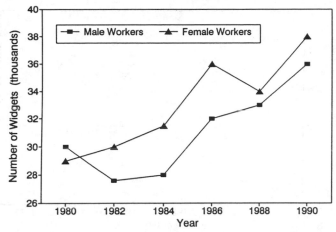

Figure 3.9. Widget Production by Sex, 1980–1990

(....); then in a vacant corner of the graph area a legend might be placed as follows:

Legend:
---Males
. . . Females

It might also be placed at the bottom of the graph in some cases, and other symbols used, as is illustrated in Figure 3-9.

As is perhaps obvious by now, any frequency distribution can be presented as a graph, usually a line graph or a bar graph, although other types may occasionally be preferred. This statement is important enough to bear repeating! Any frequency distribution can be presented graphically, yielding a picture of the way the members of the population are distributed with respect to the values along the measured dimension.

Not only can a frequency distribution be presented pictorially, but it probably should be—if not for public consumption, then simply for personal edification. Many prominent researchers insist on looking at a graph before commenting on their results—even the most experienced of us can usually profit from a visual image of the data. Those of you who must work with data should form the habit of plotting your data from time to time in order to be sure you have a clear "picture" of what is going on.

As for the above rules of thumb, you should follow them wherever possible, even if you are creating a graph just for your own use—you can always kid yourself if you want to, but you should know better when you do!

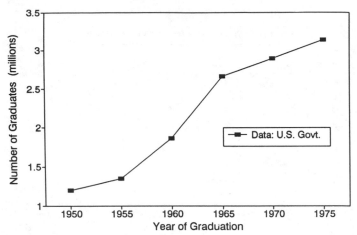

Figure 3.10. High School Graduates by Year

d. Other plots. In passing, I should point out that there are many kinds of plots that can be made other than frequency plots, although we will not discuss them in detail. For example, you could plot Number of High School Graduates by Year of Graduation (as in Figure 3-10), income by occupation, or household pets by size of household, or other information. Indeed, the process need not be confined to two-variable plots, although things get sticky rapidly when we try to get too much into a single graph.

Although we have talked so far only about frequency distributions based on a single measurement dimension, you should be aware that wherever

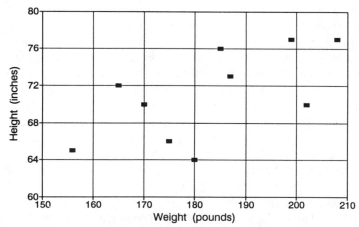

Figure 3.11. Joint Frequency Distribution ($N = 10$)

the items in the population have been subjected to more than one measurement, *joint* frequency distributions are possible. To illustrate, if we measured a group of people for both height and weight, we could develop a frequency distribution on either, or on both together, jointly. (See Figure 3-11 for a small example.)

The joint distribution of two such measures is called a "bivariate" (two-variable) distribution. Theoretically, joint distributions could be constructed for three or more measures, but the complexity of more than two rises so rapidly that we rarely consider more than two together at a time, at least for plotting and other general purposes. We will talk more about joint distributions in a later chapter.

You will note that we have now reached a point where we can arrange almost any distribution in a fashion so that the major trends in the way the observations distribute themselves along the measurement dimension can be easily seen and comprehended.

C. Some Common Distributions

There are other questions that we can answer, using statistical techniques, about our basket of numbers, and these are treated in the following chapters. However, before going on, it seems useful to present a number of standard, or typical, distributions, and to comment on the information contained in them.

I will not try to make up the actual numerical distributions in each case, but will show you the typical pictures (graphs) that would arise if such distributions were to be plotted. Remember that, in some cases, I am showing you a somewhat idealized version (graphs plotted from real data would show more irregularities).

Distributions are grouped and classified mostly by their shapes, and I will present and discuss some of the more common ones. To some degree, this is an exercise in vocabulary building—you may not wish to know some of the terms offered here, but at least you will be able to look them up if the occasion arises.

1. Symmetrical distributions

One of the classes of distributions is called "symmetrical." A symmetrical distribution is any distribution for which the left side of its graph is a mirror image of the right. Suppose you ran a vertical line up and down through the middle of the graph; then you began to fold the graph using that line as a fold line. Would the two halves match up? If so, the graph is symmetrical. (This discussion assumes that the distribution has been graphed in the standard manner—with the score scale on the horizontal axis and the frequency count on the vertical axis.)

Although the first four typical graphs I am going to show you differ drastically from each other, they all have the quality of bilateral symmetry. (The bilateral part is a fancy way of saying side-to-side, as opposed to up-and-down or top-to-bottom, symmetry.)

Another way to look at it would be: If you laid a symmetrical line graph out on a wooden board with the horizontal axis along the bottom edge of the board, and then took a saw and cut it out along the curve of the line graph, a perfectly symmetrical graph would just balance on a knife edge at the midpoint. Let's now look at each of these four types.

a. The normal curve (Figure 3-12). The first graph is called a "normal curve" in statistics; probably the name comes from the fact that it seems normal for many dimensions to be distributed this way. It has other names as well, such as "curve of (random) error," Gaussian Curve, and the "bell-shaped curve." This curve is a very specific one with a precise relationship among its characteristics defined by a complex mathematical formula (which we need not consider here). Its formula was originally worked out by the French mathematician Abraham De Moivre, and first published, in Latin, as the second supplement to his *Miscellanea Analytica* in the year 1733. Only a couple of copies of this obscure supplement remain to us, but the essence of the matter was republished a few years later by De Moivre in another work, and has since come to hold a place as one of the most significant discoveries of mathematical statistics.

The curve is derived from the relationship between frequency of occurrence and the various score values on the scale of measurement, as is true of any plotted frequency distribution. In this case, the picture is typical of

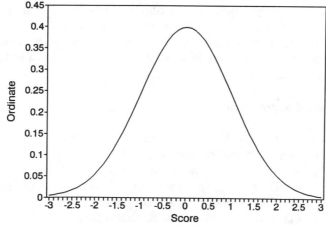

Figure 3.12. Normal Distribution

many common trait distributions in the general population, such as intelligence, athleticism, gregariousness, etc.

It is also the picture one gets if one distributes many repeated estimates of an unknown true value where the errors of estimation of the obtained scores are random (hence the title "curve of random error"). Mathematically it also represents the picture of the cumulative results of many repetitions of a long sequence of independent (unrelated), two-choice (such as Yes–No) events (for those of you who like math, equivalent to the distribution of the coefficients of a very high-powered binomial expansion). (Consult a text from the reference list for additional information about binomials.)

b. The bimodal distribution (Figure 3-13). The term "mode" means fashion or way. Did you know that the expression "a la mode" is French, and originally meant "in the manner of the fashion", i.e., in the popular way? It had nothing to do with ice cream, until ice cream became very fashionable! Anyway, statisticians call a score value that occurs very frequently a mode (i.e., it is a "popular" value).

The prefix "bi" means two; thus the term bimodal means two modes, or two very frequently occurring score values. It should be understood however that bimodal distributions are not necessarily (not even usually) symmetrical—one mode could be more popular than the other—but the one shown in Figure 3-13 has two equally popular modes, so that it is symmetrical.

True bimodal distributions are relatively rare; more often such a distribution is a clue that the population from which the data were derived might be better thought of as two somewhat different populations—

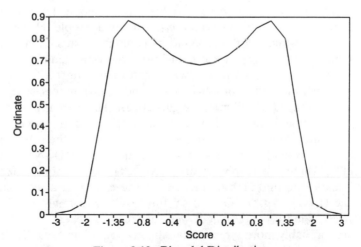

Figure 3.13. Bimodal Distribution

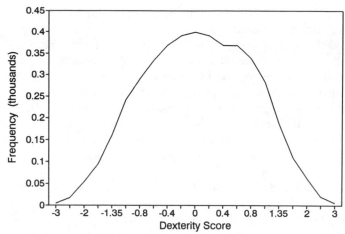

Figure 3.14. Left-Handed Dexterity

because the bimodality suggests that two groups may have been combined where they should have been studied separately.

For example, suppose we had a test of left-hand dexterity that we administered to a large, otherwise unselected group of people. If we then distributed the scores, we might find a bimodal distribution similar to the one shown in Figure 3-14.

One mode arises from the distribution of left-hand dexterity scores for people who are *left-handed*; it occurs to the right or higher side of the distribution. The other is derived from the left-hand dexterity scores for that portion of the population that is *right-handed*; it lies more to the left or lower part of the scale, because right-handed people tend to be less dexterous with their left hands than are left-handed people. However, you will note that the mode for right-handers is much stronger than that for left-handers, because there are so many more right-handed people. Overall, this bimodality suggests that if we want to study dexterity, we might wish to do separate studies of left-handers and right-handers.

c. The "U"-shaped distribution (Figure 3-15). This is really an extreme case of the bimodal distribution. It is usually found in two-choice or three-choice situations where most people prefer one or the other ends of the scale. For example, in response to the question, "Do you like the President?", where the responses are "Yes," "I Don't Know," or "No," we would expect to find high frequencies at the extremes, as compared to the middle (I don't know). Of course, the result here might not be very symmetrical, as more people might well say Yes than No.

Another, probably more symmetrical, illustration might be a distribution of a measure of Masculinity–Femininity, where, in an otherwise unse-

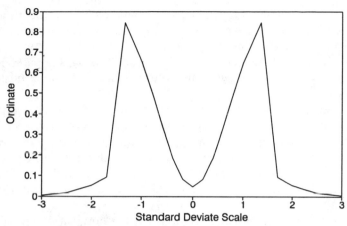

Figure 3.15. U-Shaped Distribution

lected population, we could expect a male mode at one end of the scale and an about equally strong female mode at the other, with very low incidence in the middle.

d. Rectangular distributions (Figure 3-16). Finally, the fourth graph in this group shows a graph of a "rectangular" distribution. In effect, this is a "no-modal" distribution, showing that whatever is being measured is about equally distributed among the population. Suppose you have to design a multiple-choice test, with five possible answers to each question. Other things being equal, the questions on such a test work better if the wrong answers ("misleads") are about equally popular.

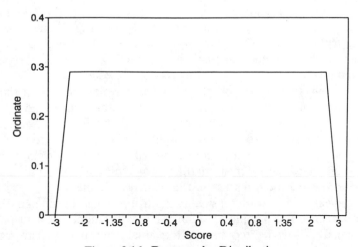

Figure 3.16. Rectangular Distribution

That is, if the right answer is chosen about 40% of the time, then, ideally, each of the four wrong answers should be chosen about 60/4 or 15% of the time. If a mislead is never chosen, it doesn't provide much information about the examinee, and is simply taking up space. In other words, you want a rectangular distribution among the four misleads to each question.

Obviously, rectangular frequency distributions are also relatively rare in terms of most measurement dimensions. In the early stages of a presidential campaign, one might find an approximation of a rectangular distribution of voter choice among six or seven candidates who were not well known, but very few phenomena are distributed naturally such that the frequencies of occurrence of "very little," "in between," and "lots" are the same.

Such a distribution with time periods on the horizontal axis characterizes situations that exhibit little or no change with time.

2. Considerations of kurtosis

I hate to bring up this word—it sort of sounds like an incurable disease. There is a subclass of symmetrical distributions that must be distinguished, and statisticians in their wisdom have chosen to tag the distinguishing characteristic of this class with the term "kurtosis." It comes from the Greek word meaning curvature. Basically, you can think of it as the quality of being thin or fat, when applied to the graph of a distribution. The reference point is a normal curve, that is, a curve is thin if it is thinner than a normal curve, fat if it is fatter than a normal curve.

If you examine Figure 3-17, you can see three distributions. The intermediate one is a normal curve presented for comparison; the "skinny" one is called "leptokurtic," again from the Greek, for "thin." The largest one is called "platykurtic," this time from the Greek for "flat" or "broad." You can easily remember this one if you think of *plat*form or *plat*eau, which tend to be flat-ish; then leptokurtic is the other one!

Distributions that are leptokurtic are ones in which most all of the items in the distribution get an intermediate or moderate score, and there are very few extreme scores in either direction. Platykurtic distributions are sort of on the way toward being rectangular ones, where there isn't a very strong middle position in the scores. Both kinds of distributions provide information that is more meaningful in combination with other kinds of information, which we'll discuss in the next two chapters.

In the meantime, consider a quality control problem in the manufacture of ball bearings. If a sample is taken for measurement, the ball bearings might equally likely be too big or too small, but if the distribution is

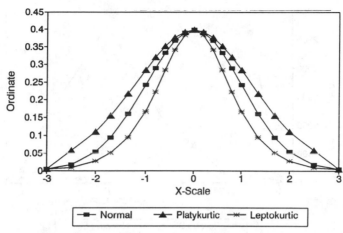

Figure 3.17. Comparisons of Kurtosis

leptokurtic around the correct size, your quality control process is quite good, and you will reject few bearings. However, if it is platykurtic, a substantial proportion of your bearings differ considerably from the correct size, and you'll be faced with a big problem.

3. Nonsymmetrical distributions

Technically, of course, this category includes all the other distributions that don't exhibit symmetry, but I won't try to discuss all the other distributions here. I will consider only two types.

a. Skewed distributions (see Figure 3-18). A distribution that looks like it tried to be symmetrical but became a little tipsy is called "skewed." It looks like it had been pushed over in one or the other direction. Such distributions are further termed either "positively skewed" or "negatively skewed." In Figure 3.18, the one on the left is positive and the one on the right is negative.

The name is derived from the position of the stretched out part or tail. You will recall that the right-hand side of any scale is assumed to be the higher, positive, or + side, and the left the lower, negative, or − side. When the tail then stretches out to the left, the distribution, or curve, is said to be negatively skewed, and when it lies on the right, positively skewed.

Skewed distributions are very common. They arise in all kinds of situations where the variable being distributed occurs somewhat disproportionately (i.e., is not "normal"). One example of positive skew would be the distribution of family income. There are very few families with no

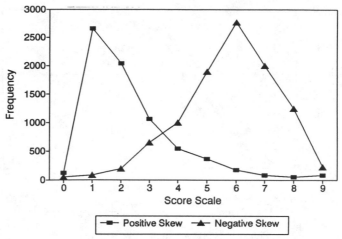

Figure 3.18. Positive and Negative Skew

income, lots with moderate income, and a decreasing number with larger and larger incomes. This distribution for the year 1975 is shown in Figure 3-19.

Figure 3-19 is worthy of some further comment. The data were taken from the U.S. Census, which used the system of unequal intervals (shown here as slashed bars) to report the data. Notice that one's impression of the shape of the distribution is strongly colored by the use of the unequal intervals. In fact, here these bars give the impression of an almost normal distribution.

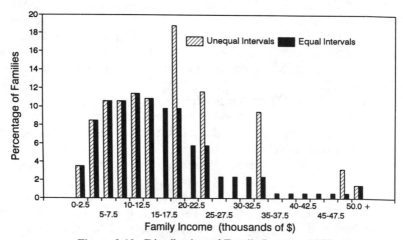

Figure 3.19. Distribution of Family Income, 1975

I have shown in the second (shaded) set of bars an estimate of what the distribution would look like if equal intervals were used. When this correct (or preferable) procedure of equal intervals on the baseline is employed, the positive skew of the graph at the upper income levels becomes highly apparent.

It is always a temptation to save space by lumping the thinly spread extremes of a distribution together into larger categories. If this is done, however, one must forever be alert to the changes in the shape of the distribution that may result, along with the different impressions, or misimpressions, of what is really going on in the data.

Negative skew arises in the opposite case, where a few cases are spread out over a wide range of low scores, lots are bunched up at a higher level, and very few are at the top. Some years ago, this situation characterized the distribution of military officers by rank. Because there were very few billets for general officers, promotions would result in a disproportionate piling up at the upper end, followed by a sharp reduction at the transition to general officer. In recent years, however, the distribution has been considerably normalized, perhaps even exhibiting a touch of positive skew. See Figure 3-20.

Skewed distributions can be a hint that something is happening to limit or "cap" the distribution. In the negatively skewed case, we refer to a "ceiling" effect. The implication is that something is interfering with the ability of the population to reach the higher score levels. Negative skew is often seen in student test performances where the test is too easy, so that both mediocre and good students get high scores (and the assumption is

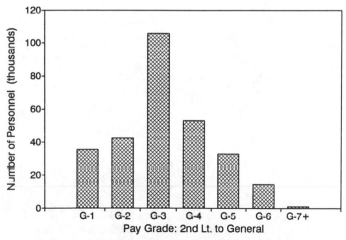

Figure 3.20. Military Personnel on Active Duty, 1988

that if the test were more appropriate in difficulty only a smaller number of very able students would get high scores and the distribution would be more normal in shape).

The converse situation exists when the skew is positive, and we say that such a distribution exhibits a "floor" effect. In the testing case, this would suggest that the test was overall too hard (it did a good job spreading out the able from the very able students at the top end, but lumped everyone else together at the lower end of the scale—neither moderate nor poor students did well).

Floor and ceiling effects are of course usually undesirable in testing people. The main purpose of testing is usually to be able to differentiate among people with respect to their measured amount of the characteristic being studied; when people are all lumped up at one or the other end of the scale, you can't tell them apart. Therefore psychometricians and test publishers generally try to avoid tests that produce these kinds of distributions.

Naturally, however, some phenomena simply exist in skewed form, and the plain fact of skewness does not necessarily demand "fixing"—at least from a statistical viewpoint. One might wish for a more normal distribution of family income, but for this result, one should see a politician, not a statistician!

b. J-Shaped distributions. So-called J-shaped distributions are really an extreme case of asymmetrical or skewed distributions. They look something like those in Figure 3-21.

The curve on the left, which is a positive J, shows a distribution that is typical of the frequency of occurrence of industrial accidents by individual

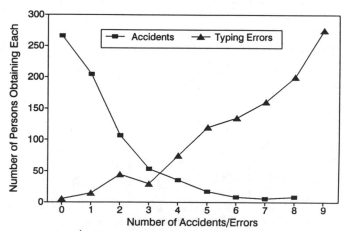

Figure 3.21. J-Shaped Graphs: Accidents/Errors

worker. It shows that the vast majority of workers have no accidents, and that a rapidly decreasing number have one and multiple accidents.

The graph on the right illustrates the converse phenomenon, one which happens more frequently at higher levels. A distribution such as this might arise from pooling the midterm typing tests in a business school. Number of errors made is put on the horizontal axis, and number of students making each number of errors (frequency) is on the vertical axis. Very few students will score no errors on this test, and the number increases as the number of errors goes higher.

There are two situations to watch for regarding J-distributions. First, if you plotted one such as the number of families with 0, 1, 2, and 3 or more TV sets, you might get a J. But, if you extended the scale farther (3, 4, 5, 6, 7, 8 . . .), you would get a skewed distribution rather than a J. (If extended, our typing errors graph would show this, too.) Thus, J-appearing distributions should be closely examined to determine the extent to which the scale has been artificially restricted. The accident distribution, above, is clearly a J; the TV example is J-like, but less so.

The other situation to watch will become clearer later, but may be summed up by saying that many statistical procedures and computations don't work too well when applied to J-distributions. This is another reason that there is no substitute for plotting and examining your data.

4. Cumulative frequency distributions

A final topic for this chapter is the cumulative frequency distribution. Fundamentally, this is simply a variation of the frequency distribution which does not provide any new or additional information about our basket of numbers beyond what is contained in the grouped frequency distributions that we have been discussing. Therefore, we will not spend much time on it. However, it is sometimes useful to examine the same thing in a somewhat different form.

The idea is to start at one end or the other of a grouped frequency distribution and progressively add the tallies from interval to interval. The usual case is to start at the low end of the scale, so that as scale values increase, so does the cumulative total of frequencies. Then the number that is listed beside each interval represents all of the frequencies *up to and including* that interval. Let's take a look at how this would work, using the distribution that I presented earlier. (See Table 3-2.)

Such a distribution permits us to make statements such as, "Some 7786 persons in our sample had 5 or fewer living siblings." Statements of this sort are sometimes useful summations, and more concise than reciting the number of tallies for each value or interval.

Table 3-2. Number of Brothers and/or Sisters Living

Number Living	Frequency (f)	Cumulative Frequency (cf)
9 or More	83	8178
8	57	8095
7	82	8038
6	170	7956
5	361	7786
4	542	7425
3	1066	6883
2	2039	5817
1	2656	3778
0	1122	1122
Total	8178	

Cumulative frequencies are even more useful in some instances in facilitating the computation of percentages for either or both of these frequency columns. This is accomplished by dividing each value in the column by the total number of persons and multiplying by 100 (to change the decimal fraction to a whole number).

For example, 7786/8178 = 0.952, which, times 100, is 95.2%. Paraphrasing the first observation made above, I can now say, "About 95% of the persons in our sample had 5 or fewer living siblings." Moving to the frequency column, if I take (2656/8178)100, I can say, "Over 32% of the persons in my sample had exactly 1 living sibling," and so forth.

As you might expect, cumulative frequency distributions can also be plotted. In this case, the score scale is on the horizontal axis and frequencies on the vertical axis, as before, but what is plotted are the cumulative tallies shown in the third column of the table.

There is a difference in where they are plotted, too. In plotting the regular frequency distribution we said that the frequency tally should be plotted at a point directly above the midpoint of the interval for that tally. Now, however, the *cumulative* frequency should be plotted at a point directly above the *upper limit* of the interval (thus emphasizing that you have to take *all* of the tallies in that interval to get the total indicated number of frequencies). (However, this instruction applies when you are cumulating upward—the usual way; if you were cumulating downward, you would plot the total to that point against the lower limit of the interval.)

Figure 3-22 shows the plot of the cumulative frequencies given in the Table 3-2. You will note that the shape of this cumulative distribution is distinctly different from that of regular frequency distributions.

The graph of a cumulative frequency distribution always takes the form of a rising curve, as no cumulative value can be less than any of its

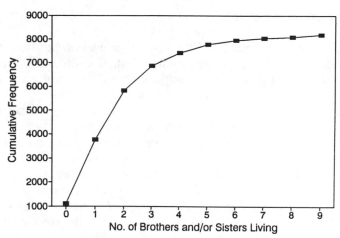

Figure 3.22. Cumulative Frequency Distribution

predecessors. Most people find it easier to study and understand graphs of the regular frequency distributions, but obviously any such distribution could also be presented as a cumulative one, and correspondingly graphed.

I will not trouble you with the cumulative form of each of the distributions we went over in the preceding sections, but there is one of greater interest than the others. As you might have suspected, it is the one for the normal curve. All of these cumulative frequency distribution plots can be called "ogives," and, of course, the one for the normal curve is called the "normal ogive." It has a nice, symmetrical "S" shape to it; see Figure 3-23.

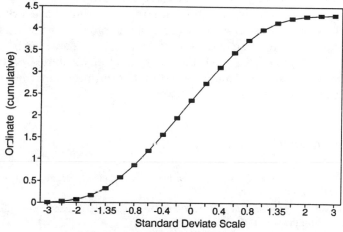

Figure 3.23. Normal Curve Ogive

III. Summary and Review

In this chapter I have presented the general problem that often confronts the statistician, or the consumer of data. That is, given a large number of observations, measurements, or numbers, how do we make any sense of them? How do we extract meaning from a mass of data?

I have suggested that the first step is to organize these data according to several well-established principles beginning with ordering them (arraying them). Because arrays can be too lengthy to comprehend all at once, the data may also be reduced by combining individual observations into an array of unique, or nonduplicative, values and showing the frequency of occurrence of each (frequency distribution). Even the frequency distribution can be reduced by forming ranges of values, or intervals of score, and tallying the individual values into these intervals. Some rules of thumb for grouping were offered.

I described a variation of the grouped frequency distribution called the cumulative frequency distribution, where the tallies are cumulated, interval by interval, so that the numbers above and below each interval become more evident. I also noted that one of the most frequent ways to study distributions was to plot them with the score scale on the horizontal axis and the frequencies of occurrence of each score on the vertical axis. Some rules of thumb for graphing were offered.

As a guide to what to expect, I presented a number of "typical" kinds of distributions in the form of graphs or plots. We noted that there are several general types of symmetrical distributions, the most important of which is the "normal" curve. I discussed several other types of distributions, both symmetrical and non-symmetrical, and offered some examples of situations which were likely to produce such graphs.

Finally, I want to reinforce the idea that studying the distribution is always important. It is easy to become submerged in the various other procedures for extracting meaning from a set of data (such as those discussed in coming chapters), but distributional attributes of the data should always be kept in mind as aids to the interpretation of not only the data, but also the other statistics as well.

IV. References

Freund, John E. and Simon, Gary A. *Modern Elementary Statistics, 8th Ed.* Englewood Cliffs, NJ: Prentice-Hall, 1992.

Howell, David C. *Fundamental Statistics for the Behavioral Sciences, 2nd Ed.* Boston: PWS-Kent Publishing Co., 1989.

Jaeger, Richard M. *Statistics: A Spectator Sport*. Beverly Hills, CA: Sage Publications, 1983.

Kotz, Samuel and Stroup, Donna F. *Educated Guesses*. New York: Marcel Dekker, 1983.

Rolph, John E. *Using Statistical Tools*. Santa Monica, CA: Rand, 1984.

Tufte, Edward R. *Visual Display of Quantitative Information*. Cheshire, CT: Graphics Press, 1983.

Tufte, Edward R. *Envisioning Information*. Cheshire, CT: Graphics Press, 1990.

Weinberg, G. H. and Schumaker, J. A. *Statistics: An Intuitive Approach, 2nd Ed*. Belmont, CA: Brooks/Cole, 1969.

Weiss, Neil A. *Elementary Statistics*. Reading, MA: Addison-Wesley, 1989.

Welkowitz, J., Ewen, R. and Cohen, J. *Introductory Statistics for the Behavioral Sciences, 3rd Ed*. New York: Academic Press, 1982.

Zeisel, Hans. *Say It with Figures, 5th Ed*. New York: Harper & Row, 1968.

4

Averages

I. Overview and Purpose

In the last chapter we were handed a basketful of numbers (observations, measurements), and the question was posed, "How do we extract meaning from such a welter of data?" It was suggested, and discussed, in that chapter that the first consideration usually is to examine how such observations are distributed with respect to the dimension under study. Distributions were discussed, and examples of various typical distributions were presented and discussed. In the present chapter, we take up the second major area of concern in the effort to extract information from such data. This has to do with answering the question, "How can we succinctly state the "typical" value for the distribution?"

It will be helpful in describing our basketful of numbers if we are able to state some "midvalue" that seems to characterize their size. Frequently we are looking for evidence of what is termed "central tendency," or, more popularly, we are looking for an "average." Thus, this chapter concerns central tendency and averages (of which there are a number of possibilities), and a discussion of the various ways to address the issue of estimating the midvalue for a distribution.

II. Central Tendency

Central tendency is one of the most important concepts in all of statistics. Just what is meant by this term? Well, statisticians have observed, and I'll bet you have too, that there always seem to be more people like other people than there are people who are very different from other people. This is true almost no matter what trait or characteristic you take (and is often true for sets of numbers that are not direct trait measurements, but are derived from manipulations and computations).

Put another way, most sets of numbers, or distributions, tend to clump up in the middle. That is, there is in most sets of data a tendency for there to be an unequal frequency of occurrence across the range of values in the distribution—fewer at the extremes, and more, proportionately, in the middle. This is what gives the familiar "bell-shaped curve" its shape—the bunching up of observations in the middle of the distribution. It also explains why true rectangular distributions are rather rare.

Thus, central tendency is the empirically observed tendency for measurements of many of the things we are interested in to bunch up in the middle (center part) of the distribution. This tendency provides a very useful way of summarizing the size of the values in a set of data. *You simply state a value on the score scale where the bunching up occurs.* Any such value is called a "measure of central tendency."

One possible area of confusion is the fact that statisticians sometimes refer to any measure of central tendency as an "average," whereas many of us have learned that an average is the result of adding up a list of things and dividing by the number of things added up. It is true that such a procedure does produce an average, or measure of central tendency. However, that particular average has an explicit name of its own: arithmetic mean. We will talk more about this later in the chapter, but for the moment remember that we are using average here in a broad sense to represent any measure of central tendency.

Average is an old concept. The word derives from the Latin, "havaria," and originated in connection with an attempt to determine a way to spread the losses in early shipping accidents (Ref.—M. J. Moroney, In The World of Mathematics, V.3, James R. Newman, Simon & Schuster, 1956.) A great many statistical techniques and theories make use of the concept of average either in their derivations or their applications.

Obviously, if all of the numbers in the set were the same, then stating just one of them would be sufficient to convey all of the information about the size of the numbers contained in the set. Of course such uniformity rarely occurs. However, to the extent that a distribution exhibits central tendency, the value that most closely characterizes that tendency is a good index of the size characteristic of the set.

If little or no central tendency is present in the distribution under consideration, then citing the score value corresponding to whatever little bunching there is conveys little or no information about the size characteristics of the set as a whole. Even so, it may still be valuable to cite what amounts to a "middle" score, while recognizing that it is not very representative of the group as a whole. (This is a very important situation, and we will return to it in the next chapter.)

As an example, suppose there were little central tendency evident in the distribution of income for each of two occupational groups. If the "middle" values of the distribution differed by $1000, this might be worth knowing, even though the middle values were not very representative of their respective groups.

Now, however, let us assume for the sake of the following discussions that our basketful of numbers does indeed exhibit a reasonable amount of central tendency. The next practical question has to do with what index or measure of central tendency one might devise in order to provide a summary value for the whole set. In other words, how do you choose a single number that best summarizes the size characteristics of the whole basketful?

Well, life is never simple. As you might expect, there are several choices, and these tend to differ among each other. Also, this is important: *Any and all averages are numbers on the score scale.* That is, an average is a *score value* that tends to typify the scores in a distribution. It is easy to lose sight of this basic fact in the details of the various choices, how to calculate them, and which are the best in what circumstances.

The right one to use depends on the nature of the data being averaged, and on the purpose of the computation. Sometimes it is incumbent upon the analyst to use more than one. The remainder of this chapter discusses these various choices and the circumstances under which they are appropriate. But, remember—they are all averages of one sort or another, and they are all in their own way measures of, or indices of, central tendency.

III. Measures of Central Tendency

In the material to follow in this chapter, I will tell you how to calculate some of these indices of central tendency. But I will not go into great detail about computations. The central purpose of these discussions is to understand the concepts involved. With the wide availability of various kinds of computational help, from hand-held calculators to sophisticated computer programs, it is much more important for the majority of you to understand the concepts and their appropriate applications than to be able to hand-calculate a lot of numbers. Of course, we will also walk through some computational examples wherever this seems to be the best way to develop and/or illustrate the concepts involved.

In much of the discussion that follows we will refer to the distribution shown in Table 4-1.

A. The Mode

We talked about the mode (symbol: Mo) in the preceding chapter. You will remember that it was defined as the most "popular" (i.e., most frequently occurring) score value in the distribution. We found that modes were useful in describing the types of distributions that we were dealing with. However, the mode may also be cited as a measure of central tendency, because in most distributions it occurs at a bunching up of scores in the midportion of the distribution. Look at the display in Table 4-1, in which I have arranged a set of scores as both a simple frequency distribution and a grouped frequency distribution.

Considering for now only the first two columns in this distribution, it is clear that the mode is 45. Looking up and down the column of frequencies, the most popular score is 45, which was the score obtained by 12 of the 131 items. It should also be noted that the mode is not a particularly good summary value for this set of scores, as there seems to be a lot of spreading out of the values in both directions. In other words, the mode is not very "strong" (pronounced) in this case.

Averages

Table 4-1. Test Data

Score Value	f	Score Interval	gf
58	1		
57	0	56–58	1
56	0		
55	2		
54	1	53–55	3
53	0		
52	5		
51	6	50–52	14
50	3		
49	8		
48	7	47–49	25
47	10		
46	10		
45	12	44–46	29
44	7		
43	6		
42	4	41–43	16
41	6		
40	8		
39	9	38–40	22
38	5		
37	4		
36	2	35–37	9
35	3		
34	2		
33	2	32–34	8
32	4		
31	3		
30	1	29–31	4
N (Total)	131		131

Some other things can be noted from examining this distribution. First, there is a tendency to bimodality. Notice that the scores become less popular from 41 to 44, but the frequency increases again for the values of 39 and 40 before dropping off toward the tail of the distribution. We might say that there is a "weak" second mode at 39–40. If you plot this distribution as a line graph, you will see the telltale second hump.

You might also observe a skewness, which is indicated by an asymmetry on one side of the distribution. In this case, there is a negative skew, since there are disproportionately many scores at the left or lower end of the

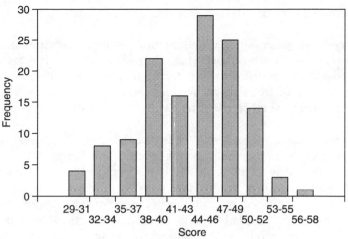

Figure 4.1. Grouped Frequency Distribution

distribution (recall that skew is identified by which side of the distribution has the "tail").

If we consider the grouped form of this distribution shown in the third and fourth columns, these points become more clear. This, as noted previously, is one of the advantages of grouping. Some of the excess detail is cleared away, and the major trends become more evident. In this case, I have used an interval of 3, reducing the score set from 29 individual scores to 10 three-step intervals. To accomplish the grouping, the frequencies for each of the three individual scores covered by an interval are added together. The grouped frequency is then listed opposite the interval.

If you are graphing, the grouped frequency is plotted opposite the *midpoint* of the interval. For illustration, Figure 4-1 shows a bar graph based upon this grouped frequency distribution.

Looking at the grouped frequencies, we would again say that the mode is 45. It is clear that the interval 44–46 is the most popular, having the greatest number of frequencies (29). In the case of grouping, the mode is always the *midpoint of the most popular interval*. Further, the secondary mode is now more clear at 39, the midpoint of the interval 38–40.

Note that this is the secondary mode even though 47–49 drew three more items than did 38–40. This is because the dip at 41–43 serves to set off the subsequent rise at 38–40, while there is no such break in the trend in the upper end of the distribution. Finally, the negative skewness is even more evident. You should plot the grouped frequency distribution so that your visual impressions can reinforce these characteristics of this distribution.

The mode is generally regarded as one of the weakest measures of central tendency. This is because it is relatively imprecise. For example, it is not at all uncommon for two or more score values to have the (same) greatest number of frequencies. Which then is the mode? They may not even be adjacent values in the array. How can one determine a compromise?

The time-honored method is to "split the difference." That is, you take the two tied values and go halfway in between. This becomes less and less satisfactory as the two values are farther apart, and becomes complex if there are more than two. In any case, such meanderings are somewhat in conflict with the definition of mode, as the result may be more central, but surely is not more "modal!" Many of us simply throw up our hands, concluding that the mode is too meaningless in such cases to be worth the effort of determining.

This situation may even arise in the case of grouped frequency distributions, with similar conclusions being drawn.

Then, of course, there is the issue of what we previously characterized as bimodal distributions, where there is a "primary" and a "secondary" mode. A bimodal distribution tells you clearly that the tendency to centrality is somewhat divided; and that you run some risk in stating that the primary mode is a good summary value for the size characteristics of the entire data set.

Finally, the mode has little relationship to any other statistics that might be computed on the distribution. Other averages, as we shall see later, can be linked to other statistics in ways that make them more valuable. In the last analysis, as a measure of central tendency, the mode is best thought of as only a crude indicator.

The most valuable aspects of the mode tend to be distributional. In considering the distribution of one's data one should keep in mind the overall importance of the shape of the distribution and what this might mean for the interpretation and analysis of the data. In this respect the mode can many times be useful.

B. The Arithmetic Mean

The term "arithmetic mean" represents what you have probably always thought of as the "average." We now require a somewhat more precise definition, because, as I mentioned above, there are various other averages in statistics as well as this one.

The term "mean" connotes in between the extremes, or intermediate. The term "arithmetic" connotes addition. Thus, the arithmetic mean is a statistic that is determined by finding an intermediate number through the process of adding up the numbers. Specifically, it is the value that results from adding up all of the numbers in a set and dividing that sum by the number of numbers added up.

Of course statisticians simplify all of the verbiage by using symbols. The symbol for the arithmetic mean is usually a literal number (a letter that stands for a particular variable), capitalized, with a line over it. For example, if we are using "x" to represent the numbers in our basketful, then X with a line over it ($\overline{X}$), pronounced X-Bar, represents the arithmetic mean of these numbers. To state the relationship in fully symbolic form, it would be $\overline{X} = \Sigma x / N$. If our numbers were weight measurements, which we had decided to represent by the literal number "w," then the relationship would be $\overline{W} = \Sigma w / N$.

Referring back to our sample frequency distribution, we can now calculate the arithmetic mean of this distribution. According to the prescription given in the preceding paragraph, we would simply add up all of the 131 scores in the distribution and divide by 131. If you do this, you will get 43.4 as the arithmetic mean, and this is the best guess we have as to the most typical score for this distribution (from the aspect of this measure of central tendency).

Of course in the actual calculations, you would not write down all 131 separate values. Where there are score values that occur repeatedly, such as 46 (which occurs 10 times), you would multiply the value times its frequency, and write down only the subtotal (here, 460). Then you would add together all of the subtotals to get the grand total (here, 5687) to be divided by the number of frequencies (131).

The arithmetic mean is only one of a number of averages which are also called means of one kind or another. But since it is by far the most commonly used of those averages, it is sometimes called simply the "mean," with the arithmetic part understood. Other means are always called by their full names, so if you see the term mean, you are justified in assuming that the arithmetic mean is what is meant.

1. The Mean as a Moment

In physics, there is a term called "moment." It is not the same moment as in "Wait a moment." Rather, it means tendency to turn or rotate about a center. Think about a child's teeter-totter, or see-saw. Here there is a center pivot with half the board on each side. As soon as a child sits on one end, the board attempts to turn on its pivot in the direction of that child. In other words, it develops a "moment."

However, this is not much fun, since there is no countermoment, and the board will only go in the direction that it is impelled by the child's weight. The child must find a playmate to sit on the other end to provide a moment in the opposite direction.

If the two children are of equal weight, they counterbalance each other, and they may bounce up and down in turn by alternately pushing with their legs. If, however, one is unduly chunky, the moment in his direction is excessive, and the moment supplied by the other child is overbalanced. Down comes the board on "Chunky's" side!

One solution is to find another playmate. However, another is simply to have the heavier one sit closer and closer to the center (pivot) until his weight at the shorter distance from the center just balances that of the lighter child at a farther distance out. (Another way to get the same effect would be to move the pivot closer to the heavier child.)

In either case, what you are doing is shortening the distance between the heavier child and the pivot, as compared to the distance between the lighter child and the pivot. From physics, you will recall that this is an example of a balanced lever system. The product of the weight times the lever-arm (the distance between the weight and the pivot point or fulcrum) on one side of the system must equal the product of the weight times the lever-arm on the other side in order for the system to balance.

I have introduced the concept of moment at this point because there is a parallel between the arithmetic mean and the teeter-totter system. If you think of the scale or baseline of the score distribution as the board of the teeter-totter, and the frequencies at each of the various score points as units of weight (call each frequency a pound if you like), then the arithmetic mean is the pivot point where the teeter-totter (distribution) just balances.

In other words the arithmetic mean is a point on the score scale where the frequencies on one side (taking into account the distance of each one from the arithmetic mean) just exactly balance the frequencies on the other side (again taking into account the distance of each one from the arithmetic mean). It is this characteristic of the arithmetic mean that leads to its description as a "moment of the distribution."

Of course in the case of the teeter-totter, or the physical lever system, the objects on either side may have different weights. This is why you deal with the product of the weight times its distance from the fulcrum or pivot. In the statistical distribution case, a frequency is a frequency is a frequency. That is, we assume that all frequencies "have equal weight." Thus, we can ignore them, and deal only with their distances from the pivot point.

In effect, then, the weights are each equal to "1," and the product of each weight times its distance is 1 times that distance, and the aggregate moment on each side is simply the sum of the distances on each side.

Let us pursue this concept a little further. Let M stand for the arithmetic mean. Then we must be concerned with the distance separating M from each observation. We will require these distances (call them "deviations from the mean"), whatever they are on one side, to be exactly counterbalanced on the other. More formally, this is equivalent to saying that the sum of the deviations on the high side of the mean must exactly equal the sum of the deviations on the low side. (In effect, just as in the physical case, all of the weight on one side times its distance from the center must equal all of the weight on the other side times its distance from the center.)

We can go further and express this characteristic of the arithmetic mean in symbolic form. Let us call our scores "xs", and remembering to treat our pluses and minuses in algebraic fashion, we would require that $\Sigma(x - M) = 0$! The sum of all of the deviations of the individual, obtained scores from the arithmetic mean is zero—regardless, for every distribution.

When you treat these differences in score value from the mean *algebraically*, the deviations of the high scores from M all get *plus* values and the deviations of the low scores from M all get *minus* values. But, we have already pointed out that the sum of these deviations on the high side must exactly equal the sum of the deviations on the low side of the arithmetic mean. Therefore, the sum of the pluses must exactly equal the sum of the minuses, and taken together in algebraic fashion, the sum of all of the deviations must exactly equal zero.

Figure 4-2 shows the distribution of Figure 4-1 with the position of the arithmetic mean on the baseline marked. If you were to cut this distributional graph out of cardboard or plywood, to scale, you would see that, just like a balance system in physics, the cutout of the distribution would balance physically with the arithmetic mean as its pivot point. That is, the clockwise moment just balances the counterclockwise moment.

It is instructive to consider what the impact of adding a single score to such a distribution is. Throwing an additional number into our basketful of numbers could have one of three effects: it might do nothing, it might increase the mean, or it might decrease the mean.

If the new score were larger than the mean, it would have a positive deviation from the mean, thus increasing the sum of the plus deviations.

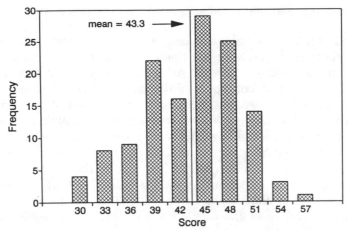

Figure 4.2. Bar Graph with Mean Score

This would cause the distribution to overbalance to the right. As we might expect intuitively, we would then have to move the pivot slightly to the right (increase the numerical value of the mean) in order to restore the balance. Thus, if a score larger than the mean is added to the distribution, the mean of the distribution must always increase to accommodate it.

If a score smaller than the mean were to be added to the distribution, the new mean would always be smaller; the sum of the negative deviations would be increased, and the mean would have to shift down slightly to restore the balance. Should the new score be exactly at the mean, there would be no change. Such a score adds nothing to either the positive deviations or to the negative deviations (its deviation is zero), and thus, no movement of the pivot point (the mean) is necessary to maintain the balance.

2. The Effect of Extreme Scores on the Mean

Note that each observation in a distribution has an influence on the mean proportional to its size (or, more properly, proportional to its deviation or distance from the mean). The more extreme the score (the more distant from the mean), the more impact it has on the mean (the more the mean must be shifted to restore the balance between the sum of the negative deviations and the sum of the positive deviations).

Recall our teeter-totter analogy. If a frequency is thought of as a pound sitting on the board, the farther out from the center you place it, the more turning force it will exert. If you want to demonstrate this to yourself, get a long stick or pole and tie a brick to the end of it. Then hold it out at arm's length with your hand placed a foot from the end with the brick. Then try it with your hand two feet from the brick, then three, and so forth. You will surely notice how much heavier the brick feels as the distance between it and the hand holding the stick increases!

This leads to a very important consideration. *As a measure of* central tendency, *the mean is greatly influenced by extreme scores.* For example, if you try to get the mean income for a group of workers, and you accidentally include just one professional basketball player in the group, your mean will be elevated to such an extent that it will be a very poor index of the typical income of these workers. (This is not true of some other measures of central tendency, as we will see later.)

To get a sense of the effect of extreme scores on the mean, let's take several simple sets of numbers. If we were asked to obtain the mean of the following five numbers: 10, 10, 10, 10, and 10, it would be obvious that the mean is 10. To calculate it, we would add up the numbers, getting 50, and divide by the number of numbers (5), with the result of 10. Suppose we had 10, 12, 14, 16, 18? It might be equally obvious to some of you that the mean here is 14, as this is a perfectly symmetrical distribution and the mean is always found at the midpoint of a perfectly symmetrical distribution. At any rate, it could be calculated the same way: $\Sigma x = 10 + 12 + 14 + 16 + 18 = 70$, and $\Sigma x / N = 70/5 = 14$, as described above.

To take a few more of these, what is the mean of 10, 11, 12, 12, 20? Adding them up gives 65 and 65/5 is 13. Notice here that the mean is higher than 4 of the 5 scores. What do you suppose is the reason? This is an illustration of the point made in the preceding paragraph about extreme scores. The value 20 is comparatively much farther away from the other 4 scores than they are from each other. Hence it has a strong tendency to drag the mean in its direction. In the case of 6, 12, 12, 13, 14, the sum is 57, and the mean is 11.4. Here the mean is smaller than all but 1 of the 5 numbers for the same reason, only here the extreme score is on the low side.

This gives rise to a very important observation: because mean means intermediate, *it can never be higher, or lower, than all of the scores in the group*).

Now look at the following: 2, 10, 11, 12, 20. The sum is 55, and the mean is 11, which is right in the middle (notice that this set of numbers is also symmetrical). However, also notice that this small set of observations is much more spread out than the last symmetrical set that we looked at three paragraphs above (where the mean was 14).

Although we could and did calculate the mean in both cases, and would probably use it to report central tendency for both of these two sets of numbers, it should be clear that this latter set exhibits considerably weaker central tendency than did the previous set (where the mean was 14). Why? Simply because the numbers as a group are not bunched up in the middle as much!

The effect of extreme values (sometimes called "outliers") on the size of the arithmetic mean is another reason why the person working with statistical data should always look at his or her distribution. Sometimes outliers can be discarded on logical grounds. Sometimes they are a clue that procedures were violated, or that errors were made. Sometimes they must be accepted. *But always* they have a strong effect on the computed arithmetic mean, and may, on occasion, suggest that some other measure is more suitable in that particular case.

In summary, remember the following points. In symmetrical distributions, the mean will be found at the midpoint of the distribution. Because of its susceptibility to extreme scores it will be found more toward the tail of a skewed distribution rather than at the modal value, as the extreme scores in such a distribution are what cause it to have a tail in the first place.

Figures 4-3 and 4-4 show the bar graphs for each of two quite skewed distributions, and the location of the mean is noted on each. You will note that the skewing diminishes the representativeness of the mean as a "typical" value for the members of the distribution.

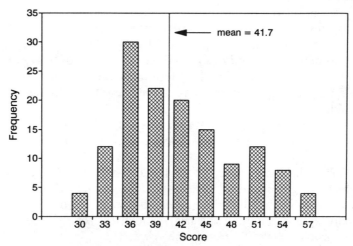

Figure 4.3. Positively Skewed Distribution

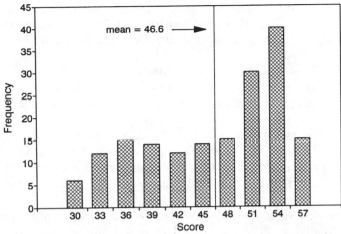

Figure 4.4. Distribution with Extreme Negative Skew

C. The Median

Another measure of central tendency is the "median" (Md). The name implies "middle," which indeed this is. (You may be familiar with the term in connection with divided highways, where the center strip of land between the opposing lanes of the highway is often called a median.)

In statistics, if you line up all of the numbers in order (make an array of them), and then count down from the top (or up from the bottom) to the middle number in the array, that number is the median. (Middle here is defined as a number such that as many of the observations are above it as are below it.)

If there is an odd number of observations in the set, this works without further ado: if the number of scores is an even number, however, there is no middle score in the sense that there is the same number of scores above it as there is below. In this case, we identify the middle *pair* in the distribution and split the difference between them, producing an artificial observation such that half of the scores are above it and half below.

Let's look at a few simple numerical examples. Suppose you have the following set of scores: 10, 14, 11, 16, 15, 18, 11. First you would rearrange them into an array: 10, 11, 11, 14, 15, 16, 18. As there are 7 scores, the middlemost one would be the 4th (with 3 scores above and 3 below). Therefore, counting up (or down) to the 4th score, the median is 14.

Let's suppose that there were an 8th score, 19. With an even number of observations, there is no middlemost score. If we took 14, there would be 3 below and 4 above; if we took 15, there would be 4 below and 3 above.

However, if we take the middlemost *pair* (14, 15), there would be exactly 3 above and 3 below, and if we now split the difference between these two, getting 14.5, and calling it the median, there are 4 scores above and 4 below.

You may say that it is too much bother to split the difference between members of the middlemost pair—any arbitrary value between the two would serve the purpose of identifying a value such that an equal number of the scores lay on either side of it. This is true to an extent. However, there are sometimes reasons to wish to make a proportional division of the score range between two values (a topic we will discuss later), and by splitting the difference between the members of the middlemost pair, we are in effect locating a median that is proportionately between the two of them.

Look once more at the set of numbers we just used, and suppose that we had added an 11 instead of a 19. This would give us: 10, 11, 11, 11, 14, 15, 16, 18. This is an even number of scores; the middlemost pair is 11, 14; splitting the difference gives 12.5 as the median. Suppose we now add a 9th score which again is 11; we have 10, 11, 11, 11, 11, 14, 15, 16, 18. The median is the 5th score in the array (with 4 above and 4 below), and the 5th score is 11. This brings out the point that *the median does not have to be a unique value*. In addition, it is clear that this distribution is heavily skewed in the positive direction.

Notice the difference in concept between the median as an index of central tendency and the arithmetic mean. Once the observations are arranged as an array, the actual values of the numbers do not enter into the calculation of the median, whereas in the case of the mean every value enters into the computation according to its exact amount (during the summation process). This is akin to using the numbers in an ordinal sense, versus using them in a cardinal sense, respectively. As a consequence, the median has no analogous aspect to the "moment" characteristic of the mean. It is called a "frequency-based" measure because, once the scores are arrayed, the median is determined simply by counting the positions in the array.

The ordinal or frequency nature of the median as compared to the cardinal or moment nature of the mean has a very important implication. *The impact of extreme scores on the median is very much less than it is on the mean*—so much so that the existence of extreme scores is one of the most important arguments for choosing to use the median over the mean as a measure of central tendency.

For example, if there were 9 scores in a distribution, and the median were 15, it would make no difference at all what the actual values of the top 4 scores were, so long as they exceeded 15. They could be 16, 16, 18,

19; or they could be 16, 34, 56, 1022; and the median would still be 15. As we have noted before, in computing the mean, on the other hand, the exact values of the scores are very important, because they are all added together in the calculation.

To put it still another way, if you add one new, additional score to a distribution, you will add exactly that much to the "Σx" which is the numerator of the mean. For the median, in contrast, you will shift the middlemost point in the distribution one step in the array, either up or down, depending on whether your additional score was above or below the old median. If there is considerable central tendency, such a shift will have little or no effect on the median value. It is for this reason that we often choose the median as the average to work with if we know that there will be some extreme scores that might act to distort our perceptions of the average of the scores.

Family income is a case in point. For most occupational groups there are usually some people who do extremely well. Using the mean projects the picture that the average income for the group is substantially higher than is true for the majority of its members, because it is pulled up by those few highly successful members of the group. A median calculation, however, gives an average income that is more typical of the bulk of the group, since in such a calculation the extremely successful count only as single observations, like everyone else, instead of in proportion to their actual incomes.

At this point it is legitimate to ask why we don't just use the median, and forget about the mean. The answer is somewhat complex. First, from our discussion of the nature of numbers, you will recall that the cardinal usage of the number system is inherently more precise than is the ordinal. This is because cardinal numbers are always considered in terms of their precise amounts, whereas ordinal numbers are treated only in terms of their positions in a distribution. Because the mean is in essence a cardinal measure, and the median an ordinal measure, it is clear that the mean is a more powerful (precise) statistic.

A second reason for preferring the mean over the median in many cases is that when we wish to consider the "second moment" of the distribution (see the next chapter), we find that the mean has a natural companion measure describing the variability or dispersion among the scores. This measure is also a cardinal measure, as compared to the companion measure to the median, which is again an ordinally based measure. So, again, we prefer the more precise of the two when other considerations are not overriding.

What is the relationship if any between the mean and the median? For practice, you might want to go back and calculate the medians for those simple number sets I used to illustrate the mean, and the means for those I used to illustrate the median. If you do, you will find some simple and intuitively obvious facts.

First, in symmetrical distributions, the mean and the median will coincide (of course, not exactly, unless the distribution is perfectly symmetrical). This is because in a symmetrical distribution, there is just as much above as below the midpoint on the distribution—but that is what it takes to make the moments on one side equal the moments on the other (the mean), and also to get half the observations on each side (the median). [Except in unusual distributions, such as the U-shaped ones, symmetry also results in the mode coinciding with the mean and the median as well.]

In the nonsymmetrical distributions, the mean and the median do not coincide. Why? Because the mean is always pulled in the direction of the skew (tail) by the existence of the extreme (at least relatively) scores that produced the skewness (asymmetry) in the first place. In fact, this difference between the mean and the median of a distribution is so fundamental to skewness that the most common index of skewness in use is based upon it. (Measures of skewness, i.e., attempts to quantify skewness, are beyond the purposes of this book, and thus will not be treated further here. The reader is referred to standard texts on statistics listed in the references for further treatment of this topic.)

To illustrate the relative positions of the mean and median, Figures 4-5 and 4-6 repeat the previous two figures, but with the Medians as well as the means indicated.

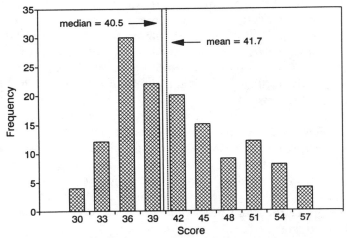

Figure 4.5. Positive Skew, Showing Mean and Median

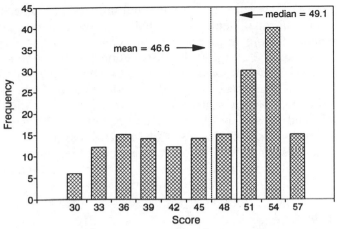

Figure 4.6. Negative Skew, Showing Mean and Median

You will notice that in Figure 4-5, which exhibits a mild positive skew, the mean is as expected slightly higher than the median. However, the difference is not great, and either of these two measures of central tendency would give about the same impression when used as typical values.

In Figure 4-6, however, the skew (negative this time) is more pronounced. Correspondingly, the difference between the two measures is considerable, again with the mean being drawn in the direction of the tail of the distribution. Here somewhat different impressions of the typical value for the distribution would be obtained.

D. The Mean and the Median from Grouped Frequency Distributions

So far, we have considered only the mean and the median as developed from standard, ungrouped distributions. Although, again, it is not my purpose to focus on calculations, it is probably desirable for us to look at these measures when they are derived from the grouped frequency distribution—just to see, and to be sure, that what we are talking about is no mystery, but rather only a simple variation of the basic processes we have discussed above. An additional benefit is to let you know what your computer is really doing when it produces these kinds of analyses.

Statistical calculations based on a grouped frequency distribution depend on two basic assumptions. These assumptions are necessary because, once the observations have been submerged into the grouping intervals, they lose their individual identities. Normally, we would be presented only with a column of score intervals and a column of their corresponding frequencies (columns 3 and 4 in the distribution that was given in Table 4-1), and asked what would be a score value to typify this distribution.

Under such circumstances, it would be impossible to know what the original scores had been for any interval. For example, for the interval 32–34, we would know only that the 8 observations in that interval had been some combination of 32s, 33s, and 34s. This loss of ability to track the individual observations almost always requires the following two assumptions in dealing statistically with the problem of calculating means and medians from the grouped distribution:

1. Assumption no. 1

The dimension underlying the score scale is a continuum. This means that we imagine that any conceivable fractional score is conceptually possible, even though we know that only whole number scores were gathered. The purpose of this assumption is to avoid the logical inconsistency of gaps in the concept being measured, which otherwise would exist between each of the intervals. The reflection of this assumption (mentioned in passing earlier) is that each interval is thought of as running half way to the next interval, even though such intermediate scores are not a part of the measurement.

If you remember, this device was mentioned as the way we ensure that there is no conflict about which interval a score belongs to. For example, the interval of 38–40 is thought of as running from 37.5 to 40.5; the next interval (41–43) runs from 40.5 to 43.5; and the next from 43.5 to 46.5; then 46.5 to 49.5; and so on. Thus, not only is the dimension made continuous, without gaps, but of course any whole numbered score can be assigned to an interval unambiguously.

2. Assumption no. 2

No matter how many observations there are in an interval, they are all considered to be uniformly spread throughout the entire interval. (Imagine the interval 37.5–40.5 as a piece of bread, and the 22 frequencies found there as 22 pats of butter—then spread them all together evenly on the bread. For the interval 52.5–55.5 there are only 3 pats of butter—no mind, spread them all together evenly over the piece of bread.)

This assumption has two virtues. First, it means that the total number of observations in the interval can be treated as though they were all concentrated at the midpoint of the interval. This is important to certain calculations. This treatment is analogous to the physical principle that when the weight of an object is uniformly distributed throughout its structure, that weight *acts* as though it were fully concentrated at the midpoint of the structure (the concept of "center of gravity").

Second, this assumption permits us to estimate proportions of the interval based on proportions of the frequencies in the interval. To push our bread and butter example a little further, the assumption of uniform distribution permits us to estimate the proportion of the slice of bread we have by knowing what fraction of the pats of butter we have. If the frequencies were not uniformly distributed throughout the interval, this process would not be possible. (We will discuss this process in more detail later in this chapter in the section titled "Interpolation.")

3. The mean from a grouped frequency distribution

Thus, to calculate the mean of a grouped frequency distribution, one has only to imagine that each of the frequencies in an interval represents an individual score *the value of which is the midpoint of the interval* (by assumption 2, above). Then, if all of these values are added up, and the sum divided by the number of values, we will have the mean, as usual.

Of course, if, as was the case in the sample distribution presented in Table 4-1, and reproduced in Table 4-2, you have 22 scores in a single interval (here, the interval 38–40), you don't write down the midpoint (39) 22 times. You simply multiply it by 22, and write down the result as the subtotal representing that interval. After doing this for each interval, you add up all of the subtotals (giving you 5676 in the case of the sample distribution), and then divide by the total N(umber) of frequencies (131). Table 4-2 shows the sample distribution (only the grouped form), with these calculations entered in. Thus, the mean from the grouped frequencies is 43.3.

Note that this mean differs slightly from that which was obtained from the ungrouped distribution (43.4). This is because the process of submerging the exact individual scores into the intervals, where they lose their exact values and are replaced by the interval Midpoints, inevitably entails some imprecision arising from the substitution.

4. Interpolating to find a median

I mentioned above that Assumption No. 2, viz., the uniform spread of all of the interval's frequencies across the entire interval, had a second virtue in the process of "interpolation". (This is the reason we don't simply

Table 4-2. Test Data

Score Interval		gf	Midpoint	gf × Midpt
56–58		1	57	57
53–55		3	54	162
50–52		14	51	714
47–49		25	48	1200
44–46		29	45	1305
41–43		16	42	672
38–40		22	39	858
35–37		9	36	324
32–34		8	33	264
29–31		4	30	120
	N (Total)	131		5676

Mean $= \Sigma x/N = 5676/131 = 43.3$

assume that all of the frequencies are at the Midpoint in the first place.) Those of you who remember some of the details of your math education will remember interpolation, but others may wish to review the material on interpolation in the accompanying box.

I noted earlier that many statistical techniques and theories incorporate the idea of averages, making this concept a very important one in statistics. Another one of the pervasive ideas in both statistics and math is "proportionality." Interpolation is based upon the notion of proportionality. [We might well have discussed these notions in Chapter 1 as aspects of the manipulation of numbers, but I felt that they would fit better here where they may be immediately applied.]

A proportion simply states the equality of two fractions, e.g., $3/6 = 4/8$. In "proportion language," this is read: 3 is to 6 as 4 is to 8. You will see in some books the same proportionality written in the following manner: $3:6::4:8$, which is read the same way. Although any two fractions which are equal can be stated as a proportionality, we don't usually do this unless we believe that there is an underlying scale(s), and we wish to emphasize the relationships among various parts of the scale(s).

Almost everybody is familiar with the concept of "splitting the difference." This concept is a proportionality concept based upon the notion of fairness embodied in "half to you and half to me." The proportionality notion comes from the fact that when the disputed amount (e.g., the difference between the seller's price and the buyer's offer) is taken, the proportion of the disputed amount received by each party should be equal, viz., $1/2$.

Other fairness notions include things like, "if there are 5 children in the estate, then it ought to be divided proportionally, i.e., $1/5$ to each." Or, "if 3 men contributed different amounts to a business enterprise, then they should receive the rewards in proportion to their contributions" (e.g., A, B, and C put in 50%, 30% and 20%, respectively, of the capital necessary to buy a piece of property; when it was sold for a profit of $10,000, this should be distributed $5000, $3000, $2000, respectively).

The word interpolation means roughly "put in between." It refers to a process in which we can estimate a point in between two known points by using a known relationship between this scale and another. Fundamentally, the process depends on the idea that the difference between any two points on the first scale is proportionate to the corresponding difference on the other.

We will deal only with the case where the relationship between the two scales is one-to-one (linear). Look at the following example:

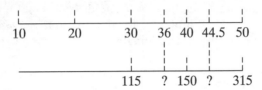

Let's consider the first "?." The basic question reflected here is that, given these two scales, and given that we have the number "36" on one, what is the equivalent number on the other? To find the answer, we assume linear proportionality between the two scales and argue as follows: If 30 on the first scale is equivalent to 115 on the second, and, similarly, 40 is equivalent to 150, the best way to estimate the equivalent for 36 is to use a proportion from the first scale and assume that the same proportion holds for the second scale.

Thus, if 36 is 6/10ths of the way from 30 to 40, then the required number (?) must be 6/10ths of the way from 115 to 150! Well, it is 35 units the whole way (150 − 115), and 6/10ths of that is 21, so the required number must be 136 (115 + 21).

Taking a second example at the second question mark, note that the target scale does not itself have to be linear from beginning to end, we just assume linearity within the interval in question. If we want the equivalent number for 44.5, we ascertain that this is 4.5/10ths of the interval on the first scale (44.5 − 40)/(50 − 40). On the target scale, the full interval is 165 (315 − 150). We want 4.5/10ths of it, which is (4.5/10)(165), or 74.25. Now we add the required portion of the target interval (74.25) to the scale position where that target interval starts (150) in order to find our place on the scale; 150 + 74.25 = 224.25, which then is the equivalent to 44.5 that we wanted to find.

Although this may seem a little complex, remember that you are simply using the proportionate position on the first scale to estimate the equivalent position on the target scale. One way to get the hang of this is to draw two scales of your own, like the ones above. Cut them with vertical lines to mark off equivalent positions. Then use one to estimate the other and see if your numbers don't come out at approximately the same place on the scale that your vertical line cuts it. After some practice with this, the interpolation-based-on-proportionality idea should become intuitively so clear that the technique will seem simple.

I have reproduced in Table 4-3 the grouped frequency distribution I set forth previously. However, instead of the columns involving the midpoint, I have substituted a column headed "cf," for cumulative frequency. In this column I have simply recorded the cumulation of the frequencies, interval by interval, beginning at the bottom of the distribution. Of course the last, or top, entry is the grand total (N) for the distribution, since the addition of the frequencies in the final interval should bring the cumulation to the grand total.

We will now use this sample distribution once more to illustrate the process involved in finding the median from a grouped frequency distribution. You will remember that the median is to be the "middlemost" score in the distribution—one such that as many scores are above it as are below.

In the ungrouped case, it was easy to determine the median, since with an odd number of scores it was simply whichever one had the same number of scores above and below. In even numbered cases, we had to identify the middlemost *pair*, and then split the difference between them,

Table 4-3. Test Data

Score Interval	gf	cf
56–58	1	131
53–55	3	130
50–52	14	127
47–49	25	113
44–46	29	88
41–43	16	59
38–40	22	43
35–37	9	21
32–34	8	12
29–31	4	4
N (Total)	131	

in order to have a plausible point on the score scale that met the requirement of half the scores above and half below.

Now, in the grouped case, we can no longer identify individual scores, so that we must *estimate* a point that plausibly meets the requirement of half the scores above and below.

If, instead of estimating, we simply settled for an interval, that is, the one that contained the 50% point, we could state a median interval. But, such a procedure produces a median that is very imprecise—in the above distribution it would be a 3-point range. If the grouping interval had been larger, the median would be even less precise.

So, the preferred process is to first identify the median interval (the one that contains the 50%-of-the-scores-point), and then use the process of interpolation to estimate how far into that interval one would have to go in order to arrive exactly at the 50% point.

To do the estimation, it is first necessary to find the median interval, or the interval that contains the score where 50% of the observations are above or below. (*Note*: Conceptually it makes no difference whether we count up from the bottom or down from the top to find the 50% point. However, by convention, we will count up from the bottom, and as a consequence the median is technically defined as: *That point on the score scale such that 50% of the scores lie below it*.)

We first need to know just how many scores it takes to get to the 50% point. In our sample case, there are 131 in all, and 50% of 131 is 65.5. So, we want to find a score value where 65.5 scores lie below that point.

Looking back at the distribution, notice that the column of cumulative frequencies, beginning at the bottom of the distribution, can facilitate the process of counting up from the bottom. Looking at this column, we have 59 of the required 65.5 cases up through the interval 41–43, but 88 (far too many) if we proceed through the next interval (44–46). We can designate the interval 44–46 as the median interval, because it contains the 50% point we want, 65.5 scores.

As I said before, if we were to be content with the imprecise result of simply specifying an interval we would be done. However, in order to get a more precise value for the median, we need to perform a procedure to estimate just how far up into the median interval we would have to go *on the score scale* in order to get precisely 65.5 of the observations.

At this point, the wisdom of assumption 2, above, should become apparent. If the 29 observations in the median interval are uniformly spread over the entire 3-point range of the interval, like so many pats of butter on a piece of bread, then we need merely advance *proportionately far* along the score scale in order to add the 6.5 frequencies (65.5–59) yet needed to reach the point on the score scale where 50% of the frequencies (scores) lie below it.

In other words, we will use the distribution of frequencies to interpolate on the score scale.

What proportion of the median interval is required? Well, it contains 29 scores, and we need 6.5 of them. Thus, $6.5/29 = 0.224$ or 22.4% of the interval on the frequency side. But this proportion must be applied to the score scale side, where the interval is 3 score points, not 29 frequencies. Applying the proportion, that is, multiplying 3 times .224, we find that the amount of the median interval needed *on the score scale side* is .67. Thus, on a proportionate basis, we must go into the median interval 6.5 frequencies (.224 of the interval) on the frequency side, and the *same* proportion, .224 of the interval, or .67 score points on the score scale side, to reach the median position.

Finally, the .67 score points must be located in terms of the score scale. It is only the piece of the median interval required. Now, remembering our discussion of assumption 1, that the intervals must be thought of as beginning half way between the upper end of one and the lower end of the next, we know that the median interval actually begins at 43.5 on the score scale. So, if we start with 43.5, and add the piece that we have just determined by interpolation, we have $43.5 + .67$, which equals 44.17, or, rounded, 44.2. It is this value, 44.2, which is our best estimate for the median, from these grouped frequency data.

Let's go over this again, while looking at the critical section of the distribution in the same way that we did for the example on interpolation above:

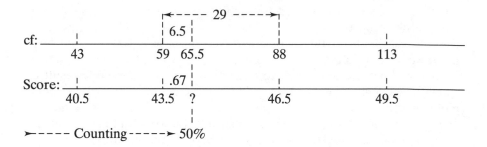

Here I have set the problem up as an interpolation problem, showing only the critical portion of the scale where we want to find the 50% point on the score scale (?). The illustration shows the corresponding points on the cumulative frequency scale (cf) and the score scale. To restate, counting up from the low end of the distribution, you get exactly 43 frequencies up through the score point 40.5; going on, you get a total of 59 when you have reached the score point 43.5, and 88 when you have gotten to the score point 46.5, and so on. Refer back to the distribution to see where these values came from.

Because the median is defined as the score point below which there are 50% of the frequencies in the distribution, we determine how many frequencies must lie below the median score point by taking 50% of the total number of frequencies (observations) in the distribution: $(.50)(131) = 65.5$; I have shown this point on the frequency scale, labeled as the 50% point.

What we wish to do now is to interpolate for this 50% point on the score scale (shown as a "?"). As before, in an interpolation problem you use a known proportion on one scale to estimate a corresponding point in the same proportion on the other scale. Here, it is clear that the required point on the cf scale, 65.5, is 6.5 observations into the median interval of 29 observations, and the proportion of the interval required is 6.5/29ths of the interval.

To find the corresponding point on the score scale, we will need the same proportion of *its* median interval of 3 score points, or 6.5/29ths of 3. Doing the arithmetic, we have $(.224)(3) = .67$ score points. Finally, in order to connect up to the scale itself, we add this piece of the median interval to what we had before, 43.5. Again, doing the arithmetic, the median point on the score scale is 43.5 plus the fraction of the interval, .67, which equals 44.17, rounding to 44.2.

Now, I am not going to give you a formula for doing a median. The problem with formulas (formulae, if you like Latin) is simply that they encourage people to apply them without first thinking of what it is they are doing. You should go over the above examples until you understand what we are doing here.

The median so derived is entirely consistent conceptually with the median we discussed in dealing with ungrouped data. It is still based on the idea of finding a score point in the middle of the distribution, such that half of the observations in the array lie above that point, and half below. The only added factor is that grouping the scores makes it impossible to identify which individual score that is, and so, we are forced to estimate, by interpolation, an artificial score that meets the requirement of half the scores above and half below.

We have now dealt with the three most common "positional" measures. All are called "averages," or measures of central tendency. Of the three, the mode is the simplest, and the arithmetic mean the most important, both practically and theoretically. However, the median, because of the fact that it is a member of the frequency system, called percentiles (which we will discuss at more length in the next chapter), also finds wide application in many situations.

E. The Quadratic Mean

I don't mean to be mean, but I have the means without going beyond my means of presenting to you yet another Mean—which is no mean trick! [Who says English is easy??]

The word "quadratic" is derived from Latin and other ancient languages, and means simply "square." Although the quadratic mean itself is rarely referred to by name, it appears in a highly important form in statistics, which is why I bring it in here. (However, the idea of a quadratic mean can be applied to any set of data.) We will discuss its important application to statistics in some detail in the next chapter.

Clearly then the accepted usage of the term quadratic is in connection with squares. What then would a quadratic mean be but a mean based on the squares of the original values. In any case *the QM is defined as the square root of the arithmetic mean of the squared numbers*.

To look at an example, suppose we have the following data set: 0, 2, 5, 6, 8. To get the QM, we would first square each value, which would give us 0, 4, 25, 36, 64; then we would get the arithmetic mean of these five squares. Summing, we have 129, which is divided by 5, or $129/5 = 25.8$, which is the arithmetic mean of the *squares*. Finally, taking the square root of 25.8 yields 5.1, which is the QM!

Notice that the QM is considerably higher than the arithmetic mean of the original values (4.2). This will generally be the case, because the difference between a number and its square increases exponentially as the number gets larger. However, the QM is a midvalue of sorts for the set.

It may be instructive to consider why one might wish to use a quadratic mean. Suppose one had some negative numbers in the set, but wished to get some idea of the mean size of the numbers irrespective of sign. There are two ways to do this. The first is to pretend that the negative signs don't exist, and to treat all of the numbers as positives, and to get the arithmetic mean. This is acceptable for some purposes, but in other cases, where the results might be applied to other situations or theory, this mean is false. It depends on a mathematical "lie" in that some values were altered (minus signs were dropped).

The other way to do this is to compute the quadratic mean, since in this process the minus signs disappear mathematically. Recall that a square is the result of multiplying a number by itself, e.g., $(-2)(-2)$. But algebra tells us that when a minus value is multiplied by another minus value, the result is positive. Therefore, as we square all of the numbers to compute the QM, all of the values become positive, and the problem of negative values disappears.

Lets take an example; suppose we have the number set: $-2, -1, 2, 4, 6$. The squares are: 4, 1, 4, 16, 36. Getting the arithmetic mean of the squares gives $61/5 = 12.2$, and taking the square root gives 3.5 as the QM.

One more thought on the subject of the quadratic mean: why bother to take the square root in the final step? The answer is simple. In brief, the original set of numbers usually results from a measurement of some type where there are units of some sort attached to the values. When you square the numbers (or multiply them together), you do the same thing to the units as well.

For example, suppose you measure your room for a carpet. If it is 3 yards on one side and 5 yards on another, you will need $(3y)(5y) = 15y^2$ or 15 *square yards* of carpet. Thus, in the example above, the 12.2 is 12.2 *square units* (whatever the unit happens to be). By taking the square root at the end we take the square root of the squared units as well, and thus get back to the original scale and units that we started with.

IV. Means and "Progressions"

In the preceding discussion, I have emphasized the idea of central tendency in connection with finding a value in the middle of the distribution that can be used to represent those individual values in the group. Some central tendency must be present in the distribution for the "middle" value computed by any of our techniques to have much utility as a representative, or typical, value for the group as a whole.

Since very many of the sets of numbers that we statisticians frequently encounter do indeed exhibit considerable central tendency, this requirement usually presents little difficulty. Because of this, the three measures of central tendency first discussed above are by far the most commonly used and frequently reported.

However, there are several other averages, all different forms of the mean, which deserve our attention before we close this chapter on measures of central tendency. They are applicable in rather specialized situations, but important to use when those situations arise. A full treat ment would take us somewhat far afield, so I will simply summarize this material here.

The emphasis with respect to these other means tends to be more on the idea of "middlemost," rather than representativeness or central tendency. In other words, if you have a sequence of numbers, what is the best or most appropriate way to identify the middle of the sequence? In order to have some idea of the situations to which I refer here, it is necessary to consider the notion of "progressions."

We say that we have a progression when the data points, or observations form a sequence (such as successive points in time), and there is a mathematical regularity that links the numbers in the sequence. In other words, if you know one of the data elements in the series, and you know the nature of the regularity, you can use these two facts to calculate the other members of the series of numbers. (We ought to note in general, *there is nothing to prevent fractions and decimals in any of these progressions as long as the basic rule of a regular relationship between adjacent terms is preserved.*)

The simplest of these cases is called an arithmetic progression or arithmetic series. We will start here, since this involves the by-now-familiar arithmetic mean.

A. The Arithmetic Progression and the Arithmetic Mean

The numbers 100, 200, 300, 400, 500 form an arithmetic series or progression. They might arise in a situation where you were squirreling away $100 each month, and they would represent your cash stash at the end of succeeding months. Inferring from this example, you note that each number in the series consists of the preceding number plus 100. Thus, an arithmetic series is comprised of a series of numbers in which adjacent numbers differ from each other by the addition (or subtraction) of the *same amount*; a regular increment, here $100.

If plotted across time (here, month by month), the plot would be a straight line with all points on the line. Future values are perfectly predictable, and can be calculated simply by applying the regular increment. For example, we know that the next two members of the series are 600 and 700 (500 + 100 and 600 + 100). Another example might be the series: 14, 12, 10, 8, 6, 4. Again, each number differs from its neighbor by a regular amount, here −2.

Given a set of numbers such as this, we might be asked to find a "centerpoint" for the set. For this purpose, the most valuable characteristic of a mean would be that whatever is computed as the mean should actually be a member of the data set, or at least on the plot line that one gets when a line graph through the points is constructed.

Let's look at the arithmetic mean for the two examples above. In the first we have 100 + 200 + 300 + 400 + 500 over 5, or 1500/5 = 300. Clearly, the arithmetic mean is appropriate for this arithmetic series, since 300 is in the set, and, since the group is symmetrical, is clearly the middle value.

To confirm our deduction that the arithmetic mean is appropriate to use to find the middle in this situation, the second data set mean is: 14 + 12 + 10 + 8 + 6 + 4 over 6, or 54/6 = 9. Although 9 is not a member of the set, it is clearly in the middle of the set (exactly), and if you plot the set across time, you will see that the position of 9, if it were in the set, would be exactly on the plot line.

From all of this, we conclude that the arithmetic mean is the most appropriate mean to use (there are other possibilities, as we shall shortly see) with data in the form of an arithmetic series. Several other forms of progressions are sometimes encountered, and, not surprisingly, there are forms of the mean that are peculiarly suited to calculating the centrality of such data sets. We will consider next the geometric series.

B. The Geometric Series and the Geometric Mean

As I have stated, the choice of the various measures of central tendency to be used with a particular set of data depends very much on the nature of the data and the dimension from which they were drawn. The arithmetic mean is adequate for many purposes, but in the case of certain progressions is misleading. Figure 4-7 provides an example.

Figure 4-7 shows a geometric curve, as contrasted to the straight line of an arithmetic progression. Such a curve results when the data form what is known as a geometric progression.

An example would be a growth curve for a self-propagating item, such as population growth. In such cases, the arithmetic mean tends to be too high (or too low, if the curve is flattening instead of growing).

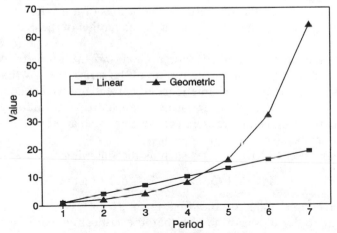

Figure 4.7. Linear vs. Geometric Plot

As in the arithmetic progression, each successive observation bears the same, regular relationship to its predecessor—but here each data point bears a *constant ratio* (rather than an additive increment) to the preceding one.

For example, the series 2, 6, 18, 54 is a geometric progression with the constant ratio of 3. That is, $6/2 = 18/6 = 54/18 = 3$. (What would the next term in this series be? I'll tell you later on.) It should be obvious that if such data are plotted you get a curve (line graph) that increases ever more rapidly. If you wish to estimate a middle value (mean) for such a series, ideally, it should lie on the plot for the data. The arithmetic mean lies on a straight line between the end points of the data array, and thus, overshoots. (See the figure.) The next term in the series above is 162.

What is required is a geometric mean (call it "G").

If there are "n" data points in the series, G is given by multiplying them all together and taking the "nth root" of the result. In the past of course one had to resort to logarithms to carry out such involved calculations, but now a good hand-held calculator can do them for a substantial series. The formula is:

$$G = \sqrt[n]{(x_1)(x_2)(x_3)\ldots(x_n)}\,.$$

C. The Harmonic Series and the Harmonic Mean

Another progression is one that we term the "harmonic progression, or harmonic series." The word comes from both Greek and Latin, and means "musical, or suitable." It has been used in physics to describe motions that take place symmetrically on opposite sides of an equilibrium point. Such motions characterize the vibration of the strings of a musical instrument, for example, and are an approximation of the motion of the common clock pendulum.

Again, as with the arithmetic progression and the geometric progression, the task is to calculate the midvalue of the series. We saw that for the arithmetic series that the arithmetic mean correctly provides the midvalue for the series, but that it somewhat overshot the correct middle term for the geometric progression we were considering—the appropriate mean for such a series being a Geometric Mean.

Now, suppose we have a series such as the following:

$$1, 1/2, 1/3, 1/4, 1/5, 1/6, 1/7, \text{etc.}$$

Or, to clarify, think of the first term as $1/1$ (which of course is equivalent to 1). It is easy to see that the critical characteristic of this series is that the *denominators* form an arithmetic progression. Or, put more precisely, *the reciprocals (defined below) of the numbers form an arithmetic progression.* Such a series is called a harmonic progression, and the correct computation of its midvalue is called a harmonic mean. (It is no coincidence that the harmonic motions of physics also involve reciprocals, though the elucidation of these relationships is beyond the scope of our discussion.)

Let's take another look at this. First, consider the notion of a reciprocal. The reciprocal of *any* number is defined as 1 divided by that number. If the number is x, then the reciprocal is $1/x$. Notice that for fractions, such as $3/4$, taking the reciprocal is equivalent to turning the fraction over (inverting), and the reciprocal of $3/4$ is $4/3$.

Thus, the reciprocal of 16 is $1/16$, of 6 is $1/6$, of $1/2$ is $1/ 1/2$ (which reduces to 2). Remember what the slash means, it means "divided by", so, $1/ 1/2$ means 1 divided by $1/2$ or (inverting, as we do with fractional division, and multiplying) 1 times $2/1 = 2$.

If we take the reciprocals in the original progression given above, we have $1, 2, 3, 4, 5, 6, 7$ as our series. Because this is an arithmetic progression now, the midvalue is given by the arithmetic mean, or $1 + 2 + 3 + 4 + 5 + 6 + 7$ divided by 7, that is, $28/7 = 4$. Looking at the progression of reciprocals, it is clear that 4 is indeed the midvalue of the progression. However, *our original progression was comprised of the reciprocals of this progression, so we must take the reciprocal of* 4 *to get the harmonic mean, which is* $1/4$. Looking at the *original progression* it is again obvious that $1/4$ is the midvalue of the progression.

To sum up, the harmonic mean (call it H) is given by: the reciprocal of the arithmetic mean of the reciprocals of the numbers in the series. In formula notation:

$$H = 1 \bigg/ \left[\frac{\Sigma(1/x_i)}{n} \right].$$

It is worth commenting that the arithmetic mean of the original progression does NOT yield the midvalue. If we compute it we get $2.59/7 = 0.37$, which is clearly way over the correct value, 0.25. We can try the geometric mean too, and this gives 0.295, which is still too high. (Incidentally, this is generally the order of these three means when applied to such data: $H \leq G \leq M$!)

So why might one be interested in the harmonic mean, anyway? Eschewing theoretical applications, there is one set of common problems for which the harmonic mean rather than the arithmetic mean is the correct measure. *This set of problems has the general formulation of quantities that are computed by multiplying together a rate and the number of units affected by the rate.* Included in the set are "distance" problems where distance is equal to the product of rate times time; cost or sales problems where cost is equal to the product of the number of units times the price per unit; and work problems where production is equal to the product of time times the rate of production (number of units produced per unit of time). In symbolic form, these problems are of the general nature of the following:

$$a = (b)(c)$$

Let me offer just one example. If a pilot flies four trips, each of the same length (say 100 miles), but flies at different rates (say, 100, 200, 300, 400 mph, respectively), what is his average rate of travel? Using the arithmetic mean, we get $100 + 200 + 300 + 400 \div 4 = 1000/4 = 250$ mph.

However, if we work it out trip by trip, we find that the first trip takes 1 hour (100/100), the second takes 1/2 hour (100/200), the third takes 1/3 hour (100/300), and the fourth takes 1/4 hour (100/400). Thus, the total time for the four trips is 2.083 hours, and the total distance of course is 400 miles; since rate is defined as distance over time, we have $400 \div 2.083 = 192$ mph, which then is the correct average rate.

If we had simply calculated H, we would have taken the reciprocals of the rates, getting 1/100, 1/200, 1/300, 1/400; then we would get the arithmetic mean, getting $25/1200 \div 4 = 25/4800$; and then we would take the reciprocal of that (4800/25), yielding 192, which is the answer we got when we worked it out trip by trip. So, it can therefore be seen that H is the proper value in this case.

There are six possible ways to write a distance problem similar to the one above. They are:

1. *What is the mean rate over trips of the same length?
2. *What is the mean time over trips of the same length?
3. *What is the mean distance for trips of the same time?
4. What is the mean rate for trips of the same time?
5. *What is the mean distance for trips at the same rate?
6. What is the mean time for trips at the same rate?

The harmonic mean is the correct one to use in each case that I have starred. The arithmetic mean is the correct one in the other two cases. Of the cases listed above, the first one is by far the most common of the six questions, although #3 comes up sometimes.

But how do you tell which is going to be correct without working out some involved proof as we did with the four trips above? The answer is fairly simple. *If the mean sought and the factor held constant are in a reciprocal relationship with each other, the harmonic mean is the one to use.*

V. Summary and Review

In this chapter we have looked at the second major step in drawing information from a mass of data. The first was look at the distribution; the second is find the typical, average, or midvalue of the data set.

There are three commonly used measures of "central tendency," all called "averages." The mode is the most frequently occurring value; it is a crude measure. The arithmetic mean, or just mean, is the most precise of the three, being a cardinal measure obtained by adding up all of the values and dividing by the number of them, but is sometimes inappropriate because of the nature of the data under study. One common problem with the mean is the existence of extreme scores or "outliers" which affect it severely. A third measure, the median, is based largely on ordinal considerations, and is relatively insensitive to outliers. It is simply the middlemost score in the arrayed distribution.

The geometric and harmonic means were presented briefly. They are particularly appropriate in cases where the data exhibit a particular progression (i.e., are in a series where each term bears the same relationship to the adjacent terms in the series). The quadratic mean was presented and discussed because it is of particular importance in statistics (see the next Chapter).

VI. References

Elzey, Freeman F. *Elementary Statistical Techniques*. Monterey, CA: Brooks/Cole, 1985.

Spence, Janet T., et al. *Elementary Statistics, 4th Ed*. Englewood Cliffs, NJ: Prentice-Hall, 1983.

See also References in Chapters 3 and 5.

5

Variability

I. Overview and Purpose

In an earlier chapter we started with a "basketful of numbers" and began the process of determining the information concealed in those numbers.

We looked first at various ways the numbers could be arranged to show some information about themselves—distributional characteristics. Then we examined some ways of identifying and reporting "central tendency" or averages.

In the present chapter we continue our efforts to learn about our basketful of numbers by asking questions about how different the individual numbers are from each other—questions of "variability." As was the case with central tendency, several measures of variability are available. In this chapter we examine the most common of these, and discuss some of their characteristics.

II. The Nature of Variability

In our examination of averages, we were seeking measures or statistics that could be used as "typical values" to represent the numerical set of values in which we were interested. When such values are examined in the light of the distributions from which they come, it is clear that some averages better represent their distributions than do others. More to the point for our present purposes, some distributions don't lend themselves to representation by a typical value as well as do others. This is because some distributions seem to have more of a tendency to bunch up in the middle than others do.

It should come as no surprise that, in addition to the measures that purport to identify or locate central tendency (averages), there are also measures that assess the extent to which central tendency exists in the distribution in the first place. These are called "measures of variability," and this chapter presents definitions and discussions concerning several of them.

First, let's be sure that we are in agreement on terminology. *Variability can be loosely defined as the tendency of the individual members of a set of numbers to differ from one another.* Thus, it is conceptually the opposite of central tendency, which is the tendency of the individual observations in a set to be similar to one another.

If you look at the four distributions pictured in Figure 5-1, you will note that all of them have an arithmetic mean of zero, but they appear to differ significantly from each other. The difference among them lies in the extent to which the scores spread out from the center of the distribution, i.e., their differing variabilities.

The lowest of them is roughly a rectangular distribution; here the individual scores or values are about as different from each other, taken over the whole group of them, as they can be (and the measure of central

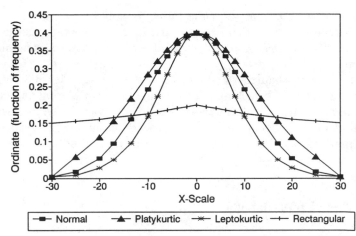

Figure 5.1. Distributions of Differing Variability

tendency, the mean, is a very poor typical value—largely because there is little central tendency in the first place).

The middle distribution of the three that group together is essentially "normal," and exhibits a modest amount of variability, as well as central tendency. The mean is a reasonably good typical value here, but there is still a considerable tendency of individual values to differ from one another.

The distribution on the inside of the normal one is one that we called "leptokurtic" earlier, and shows comparatively more central tendency and less variability—the scores tending to be bunched more closely around the mean. The more leptokurtic a distribution is, the better the mean is as a representative of the distribution.

The curve outside of the normal one is one we called "platykurtic" earlier. Here, you can see that there is relatively more variability, less central tendency, and less bunching around the mean, as compared to the normal distribution. Here the mean would be acceptable as a representative value, but distinctly less so than in the normal, or especially, the leptokurtic cases.

Measures of variability are very valuable to our analysis of the information contained in our basketful of numbers. These measures tell us something about our basketful of numbers beyond considerations of distribution and average. They enabled us to judge how representative the average value for a distribution is, and sometimes are themselves the main focus of interest.

Suppose, for example, you were a bottle manufacturer, and the above distributions represented stopper sizes from four suppliers of stoppers. Which do you suppose you might prefer to supply the stoppers for your standard bottles?

Because the essence of variability concerns the tendency of the numbers in a set to differ among themselves, it seems reasonable that the measures that have been devised to quantify this tendency depend upon analyzing the numerical differences among the numbers in some fashion. And, indeed, that is how we do it.

A. Measures of Variability

Now we will take a look at various measures of this characteristic of variability. [Note that these measures may also be known by some other names: measures of dispersion, of scatter, and even of "skedasticity." (The latter term seems to be related to the vernacular "skedaddle," which means "go away"; in this case, from each other.)]

1. The Range

The simplest measure of variability is the "range." It is defined as the score difference between the largest observation and the smallest.

Suppose we were studying ages, and we had taken a sample of 100 persons and obtained the ages of each. If we then made an array of the values, the range would simply be the difference of the two extremes in the array. In other words, if the oldest person were 79 and the youngest were 6, the range would be 73 years (and this would be a rather diverse group, agewise).

Notice that, just as was true for averages, measures of variability have the same units as do the measures themselves—in this case, years of age. This is a general characteristic of measures of variability.

To take another example, if we had measured the diameters of those stoppers we were considering for a line of bottles, and we found that the largest of the stoppers measured 0.775 inches in diameter, while the smallest measured 0.760 inches in diameter, the range would be $0.775 - 0.760 = 0.015$ inches.

Unfortunately, the beauty of simplicity is somewhat offset by the fact that the range is also somewhat simplistic! Or, more particularly, it is exceptionally vulnerable to the existence of extreme values in the number set, or what we have called "outliers." What this means is that if, for whatever reasons, some extraordinary values are included in the set of measurements, then the range can be misleadingly great.

For example, if your set of numbers consists of the IQ scores of a class of students, and by chance you have a "little Einstein" in there, the range will make it appear that the class is extremely variable in intelligence when it may in fact be rather homogeneous. Thus, even if all but one of the class has an IQ between 100 and 110, the inclusion of our "little Einstein" at 190 will change the range from 10 IQ points to 90 IQ points—giving a totally different impression of variability.

Yes, you say, but it is obvious that this is not a highly variable group, except for that one student. True, but we have the advantage of knowing all of the scores in this case. *The more usual situation is that only the summary statistics are reported*. Thus, we would read in a report that the mean was, say, 107, and the range was 90, and we would have to draw the inference that this was a highly variable group of students!

Note that this range of 90 points is *not* incorrect. It comports with the definition of the range. It simply provides a misleading idea of the variability of the *group as a whole*. As a result of this difficulty, some statisticians would disregard, or "throw out," the outlier and compute the range on the remaining observations. Unfortunately, such a procedure *is* incorrect—at least in terms of the definition of the range.

Although, in this example, such a "truncated" (cut off) range does indeed provide a better idea of the variability of this group of scores, it raises the much more difficult question of where to draw the line in general. In other words, how do you decide, consistently and with justification, when a given observation is sufficiently extreme to constitute an outlier to be discarded?

In fact, the decision to exclude a data point from statistical analysis is equivalent to saying that it is "bad" data. However, just being unusual (extra large or extra small, for instance), doesn't necessarily equate to being "bad" (in the sense of wrong, incorrect, biased, distorted, etc.). This means that the rationale for discarding a data point ought to consist of more than just the simple fact that the score is an outlier.

Rather, it ought to focus first on the "truth" of the data point; that is, an argument that the observation is flawed in some identifiable way. To be sure, outliers frequently result from, and therefore are indicators of, errors and mistakes in the various statistical processes and procedures used. But, as in the case of our "little Einstein" above, real extremes also exist, and should be dealt with in some appropriate way rather than discarded arbitrarily.

Practicing statisticians sometimes exclude outliers on the basis of their potential distorting effects on other statistics to be computed on the data.

For example, as we noted in the last chapter, the arithmetic mean is seriously impacted by extreme scores; so also are some other measures that we will consider in later chapters. The wisdom of discarding such scores, however, is always at question.

As a result of such dilemmas, many have chosen to give up the appealing simplicity of the range in favor of some other measure of variability less dependent upon extreme measures. Perhaps the coward's way out, I don't know. However, other measures have some good points of their own. Let us see.

2. Variability: Another "average" statistical process

As a perceptive reader, you have surely noticed by this time the heavy reliance that statisticians (not to mention all of the rest of us) place on the concept of average. Splitting the difference, for instance, is a time-honored American way of achieving agreement among disputants in many fields from garage sales to real estate. Suppose I want $700 for my old car, but you think it is worth only $600; is it not likely that we will compromise on $650 by "splitting the difference?"

Of course, splitting the difference is just another way of saying take the *average* (mean) of the two values. Then, too, when we had an even number of observations and wished to find a median, we took the middle *pair*, and "split the difference." Here, again, we were simply *averaging*!

In addition, with grouped frequency distributions, we assumed that all of the frequencies in an interval were evenly spread over the range of the interval so that we could treat them as so many observations at the Midpoint (which would be the *average* value under those conditions).

With such a rich tradition of averaging in statistics, it is not surprising to find that measures considered in lieu of the range also involve averaging. If you think back to the definition of variability posed at the beginning of this chapter, you will remember that variability is the tendency of the observations to differ from each other. How might we express this tendency so that the concept of "mean variability" might be developed?

On further reflection, it would appear that the major difficulty with the range as discussed above is equivalent to raising the question of how representative the range is of the distributional tendency to variability. In other words, we objected to the IQ range of 90 points where all but one of the students had a score between 100 and 110 because such a value did not seem indicative or representative of the student data as a whole.

Because the range depends on only two data points in the distribution (the largest and the smallest), perhaps the way to get a more typical or representative index of variability is to include more of the observations in its calculation. Also, since the idea of typical was related to the idea of average in the previous chapter, perhaps the way to do this is to try to define an "average" variability. In the computation of certain kinds of averages, for example, means, all of the observations are included.

a. Mean variability. The above argument leads to the conclusion that a better measure of variability than the range might be the "mean variability." The question, of course, is mean of what? What shall we take the mean of that we may call "variability?"

Variability is the tendency of the scores to differ among themselves. Hence, we should probably be looking at a mean difference of some sort as an index of variability. However, this still leaves the question of difference from what?

It turns out that, *for any given distribution, it doesn't make any difference*. Just pick any point on the score scale as a reference point and take the difference of each score from that point (algebraically, i.e., paying attention to the signs of the differences). The arithmetic mean of those differences can be used as the mean variability of that distribution. The smaller that value, the more bunched and less variable is the distribution; and the larger that value, the more spread out and variable it is. For an elaboration of this thinking, see the adjacent box.

Suppose we took a set of four numbers: 2, 6, 7, 10. If we take the difference of each of them with one of the others, we get a set of differences that we could average to get a "mean difference." However, the numerical value of this mean difference is entirely dependent on which of the numbers we chose as the "reference number." For example, if we used 2 as the reference number and subtracted it from each of the four numbers, we would get differences of 0, 4, 5, and 8, with a mean of 4 1/4. On the other hand, if we had used 7, we would get differences of -5, -1, 0, and 3, with a mean of $-3/4$; and so forth.

Because the variability in the set of numbers cannot have changed simply as a function of choosing the reference point, how can we deal with such wildly different estimates of the mean variability?

All is not lost. In truth, it doesn't really matter which reference point we choose, as long as each number is compared to the same point. (We usually do this by subtracting the reference point from the number in question.) We can demonstrate this by showing that the use of different reference points yields the same average difference, if appropriate adjustments are applied to compensate. Also, if this is true, the reference point doesn't even have to be one of the points in the data set. It can be an entirely arbitrary point anywhere on the scale.

Let's review this line of reasoning: numbers that differ from each other along a scale of measurement (dimension) would also differ from any arbitrary, but fixed, point on that dimension. For example, if the numbers 2, 6, 7, and 10 are assumed to be values along a scale running form 0 to 20, it is obvious that they differ from each other, but they also differ from any fixed point on the scale chosen at random (such as 8).

Using 8 as the reference point (even though it is not a member of the number set), the differences would be -6, -2, -1, and $+2$. The sum is -7, and the arithmetic mean would be $-7/4$.

Now, just above we took this same set of values, but a different reference point, 2. The differences were: 0, 4, 5, 8, the sum was 17, and the arithmetic mean was $17/4$. But that's not the same, you say! No, but that's because of the difference in reference points. When you equalize the two in that regard, by combining the reference point with the mean difference you have in the first instance $8 - (7/4) = 25/4$; in the second instance you have $2 + 17/4 = 25/4$ — which is the same value!

Let's confirm by taking yet a third reference point, 10: the differences are -8, -4, -3, 0; the sum is -15, and the mean is $-15/4$; when combined with the reference point, 10, the result in $-15/4 + 40/4$, or $25/4$. Again, the same result.

The problem with this, of course, is that it is difficult to develop a sense as to what is large variability and what is small when the absolute size of the mean variability fluctuates so much as a function of what the reference point chosen was, as illustrated in the first paragraph in the box.

The material in the box demonstrates that we can calculate an arithmetic mean of the differences of the numbers in our set from some fixed, but arbitrary, point. Further, it demonstrates that no matter what the reference point, we can get the same mean variability among the different values, *as long as we are dealing with the same data set*.

Based on our understanding of averages, it is reasonable to suppose that such a value can be used to "represent" the variability among the observations in a distribution. Note that such a mean difference is not immune to the effect of extreme scores (no arithmetic mean is), but it surely is less vulnerable to such scores than is the range.

The upshot of all of this is that it is possible to talk about a "mean deviation" as a measure of the variability present in a particular data set. (Deviation is the term that statisticians prefer in this context. It refers to the difference between some scale value, score, observation, or data point, and a reference point that is theoretically on the scale, but not necessarily one of the data points.)

However, intuitively, we would probably like to believe that our index of variability should not depend entirely on whatever arbitrary number was chosen for a reference point. It would be better if we could identify a "standard" reference point.

b. The "standard" reference point. As noted above, the size of any deviation (difference), and thus of the average deviation, is clearly a function of the reference point. Because situations often occur where one wishes to make some statements about different data sets, sometimes even when the numbers are derived from different scales or even different concepts, it would be useful if there were consistency in the reference points used from scale to scale.

This would make it possible to think about the variability in different data sets in a comparative way. Clearly no comparisons can be made at all if the distributions being compared employ different reference points on different scales.

Let's develop the requirement a little further. I think you can see that if the concept of mean variability is to have any real utility, it will have to be based upon reference to some point on the scale that has meaning of its own—*and* exists for scales in general.

But what should it be? A little reflection shows that if the scales are to be different from measurement to measurement and number set to number set, *there will be no way to choose an absolute value* for a reference point. Think for example of the scale of bottle stopper diameters and the scale of bottle weights, or family income, or IQ, etc. The answer of course is to *choose a relative value*—but one that has meaning in and of itself, so that it makes sense to report the average deviation of a score from a point on the scale that is easy to understand and appreciate.

We must look for a logical reference point that would have meaning in a variety of situations. Fortunately, we don't have to look too far.

The characteristics of an ideal reference point, which would bridge different distributions and different scale metrics, etc., would include meaningfulness on its own account from distribution to distribution and logical acceptability as a point of reference for describing the variability of a distribution.

The one point that is meaningful, yet relative (since it can be calculated for any distribution), is the *mean of the distribution*. Then we can talk about the average amount by which the individual values differ from the mean of those same values! Also, since this can be done for different distributions with different scales and different units, it provides a way to make some (albeit sometimes limited) comparative statements about distributions.

c. The average deviation. Thus, the second measure of variability that we want to have (the range was the first) would seem to be a mean deviation. It would be calculated by writing down the set of differences obtained by subtracting the arithmetic mean of the distribution from each of the numbers in a number set, and then calculating the arithmetic mean of those deviations.

However, there seems to be a problem here! Each time we do this—no matter what the distribution—*the answer is always zero*!

Earlier we looked at a physical analogy between the arithmetic mean and a leverage system fulcrum—it is the point where the distribution has an equal moment (turning force) to the left or right, i.e., where it would balance.

You will recall that characteristic is equivalent to saying that the scores are distributed so that the sum of the deviations from the mean on the high side is just equal to the sum of the deviations from the mean on the low side—or, algebraically, the sum of the + deviations = the sum of the − deviations, so that the sum of the two together is always zero—and, of course, zero divided by any N is also zero.

What a shame! Just when we thought we had devised a good measure of variability, we find out that it always come out zero—useless! Still, each of those deviations from the mean of the distribution is individually meaningful. The only problem is that they cancel each other out when they are all added together. What would be the situation if we simply forget that they have + and − signs attached to them?

This simple sort of "mathematical white lie" is sometimes used to salvage the mean deviation as a measure of variability. However, because it isn't really a mean when the signs are ignored, it goes by a closely related name instead: the average deviation (AD).

When the signs are omitted from numbers, we say that we are dealing with "absolute" values. This is indicated by placing the number between two vertical lines, for example, $|-6| = +6$, etc. Generally speaking, if the variable is "x," the deviation corresponding to the ith score, x_i, is d_i, and is given as: $d_i = x_i - M_x$; and, the corresponding absolute deviation, $|d_i|$ is simply the same numerical value as d_i with any negative signs dropped.

Thus, the average deviation is defined as the arithmetic mean of the absolute deviations of the numbers from the arithmetic mean of those numbers. In symbolic terms, we have: $AD = \Sigma|d|/n$, where the ds are the deviations obtained by subtracting the mean of the numbers from each of the numbers in turn, and dropping any minus signs.

The AD is not widely used. As a result of its treatment of the minus signs, it lacks some of the theoretical applications of a somewhat similar measure we'll discuss in the next section.

It is generally applicable to situations in which the interest in variability is specific to the distribution under study, and does not require further analyses beyond simply estimating the mean and the magnitude of the variability in the scores around that mean. More sophisticated analyses generally require the more theoretically sound measure alluded to earlier, as its characteristics are integral to the theory on which the more advanced techniques are based.

d. The standard deviation. The measure of choice in variability is the "standard" deviation—so called, presumably, because it is in widespread, standard usage. The standard deviation is known by many names or symbols: "SD," or sometimes "s" or "S," or sometimes a lowercase Greek sigma (σ), which is the Greek equivalent of an s. If you use the name sigma you must be sure to be clear about the difference between the lowercase sigma (σ) which is the term for the standard deviation, and a capital sigma (Σ) which is used to mean "summation." Most authors use either s or σ, and some even distinguish between the two in certain situations and formulas. Since we are not much concerned with derivations and formulas at this point, we will use SD for now.

The SD is a complex measure with some interesting properties (which are the primary reason that it is in "standard" usage). However, most of those properties are of interest in connection with some more advanced statistical ideas, particularly sampling theory, some of which we will take up later. For now, we want to consider the SD only as measure of the variability in a set of numbers, just as we did the AD. In this restricted context we would surely prefer the AD for simplicity if it were not for the way it treats negative deviations. What, then, is the SD, and how does it improve on this failing that we have noted in the AD?

e. The standard deviation as an average. We have just discussed the application of the principles of averaging in our development of the average deviation. It should then hardly be surprising to find that the SD is simply another instance of applying the principle of averaging. To be sure, it is a less familiar average then the arithmetic mean, but it is an average nonetheless.

To see how this follows, let's go back to the fundamental problem we had with the average deviation. We can illustrate it with the data shown in Table 5-1 [which are somewhat artificial in that I have arranged them for convenience so that the mean comes out even ($\Sigma x = 3600$ and $3600/80 = 45$)]. Note that this really is just for convenience—our discussion would hold for all such data sets.

You can easily see the difficulty with the AD in this example. The deviations from the mean (derived by subtracting the mean, 45, from each of the scores in turn) contain some negatives. As you can see, there are 37 scores above the mean, which have positive deviations; and, there are 31 scores below the mean, each of which has a negative deviation. Then, there are 12 scores right at the mean, which of course have zero deviation. The "fd" column shows what happens when you add to these 80 deviations (remember that fd means frequency times the value of the deviation, and is just a short cut for writing down all 80 of the deviations separately).

Now, it is clear that the 37 deviations above the mean add to $+95$, whereas the 31 deviations below the mean add to -95. Adding them all together, along with the 12 zeros at the mean, results in a total sum of the

Table 5-1. Deviation Data

Score Value	f	Deviation (d)	d²	fd	fd²
50	3	5	25	15	75
49	8	4	16	32	128
48	6	3	9	18	54
47	10	2	4	20	40
46	10 $\overline{37}$	1	1	10	10
45	12 $\overline{12}$	0	0	0 $\overline{+95}$	0
44	7	−1	1	−7	7
43	6	−2	4	−12	24
42	4	−3	9	−12	36
41	6	−4	16	−24	96
40	8 $\overline{31}$	−5	25	−40 $\overline{-95}$	200
N (Total)	80			0	670

deviations of 0: $+95 + (-95) + (12)(0) = 0$. Thus, as noted previously, as long as we define the mean difference as being an arithmetic mean (as in the AD), every distribution is going to give us a zero for the sum, and, because zero divided by any N is still zero, the mean difference will also be zero.

This may seem counter intuitive, so you should prove to yourself, using as many small number sets as necessary, that: *the arithmetic mean of the deviations from the arithmetic mean of a set of numbers is ALWAYS zero*!

Now, if the SD is, as we have said, based on the principle of averaging, just as is the AD, then how do we circumvent the problem of negative deviations and the $\Sigma d = 0$ issue? For the answer, refer to the section on the quadratic mean in the preceding chapter.

Again, the quadratic mean is the arithmetic mean *of the squares* of a set of numbers, *square-rooted* to bring the result back to the scale and units of the original set of numbers.

However, what advantage can there be in turning to such a seemingly complicated, even esoteric, notion? The quadratic mean offers a legitimate way of handling the crucial difficulty in constructing a "mean difference," which is what we want to use as a measure of the variability or tendency of the numbers in a set to differ from each other.

As we have seen previously the villains in our piece are the negative differences that reduce the computation of all arithmetic mean deviations to zero. But instead of taking the arithmetic mean of the deviations, let's look at the quadratic mean. Here is a measure that is one of the family of averages, and that starts not with the values to be averaged, but rather with the *squares* of those values. How can this be useful?

You will recall that the square of a number is the number times itself, and that when a negative number is multiplied by another negative number, the result is always a positive number. Then it follows that squaring our negative deviations, as we do in calculating the quadratic mean, is taking a negative times a negative in each case, which converts them all to positives and eliminates the "$\Sigma d = 0$ issue" entirely!

In addition to making it possible to compute a mean difference in the first place, using the quadratic mean deviation instead of the arithmetic mean deviation has the theoretical advantage of employing a mathematically more acceptable way of dealing with the negative differences that plagued our efforts to develop a measure of variability. As I have indicated before, the resulting measure, the standard deviation, has been such a felicitous result as to have become truly the standard index of variability.

Let's look back at the data presented in the last table. To calculate a SD from these data, we square each deviation and multiply it by the number of times it occurs (its frequency). Adding up the resulting fd^2 column gives a total of 670. The arithmetic mean of these squared deviations is $670/80 = 8.38$. But now to get the QM, we must take the square root, so that the quadratic mean (SD) will have the same scale of units as the original set of numbers: $\sqrt{8.38} = 2.89$ score points—which is the SD.

IMPORTANT: At this point I should hasten to note that, although correct for definitional purposes, the above result will be a little smaller than what you will get out of most calculators. This is because, for mathematical reasons that we need not consider here, in most applications we use a denominator of $N - 1$ instead of N! This has to do with a concept called "degrees of freedom," and we will talk a little about it in a later chapter. I will not trouble you with it for now, except to note it and emphasize that it is numerically of little importance unless you are dealing with very small Ns. (For example, using $N - 1$ in the above instance yields 2.91 instead of 2.89.)

In the statistician's notation, remembering that a deviation consists of subtracting the arithmetic mean from a score [$d = (x - M_x)$]:

$$SD_x = \sqrt{\left[\Sigma(x - M_x)^2/N - 1\right]}$$

How do we interpret the standard (quadratic mean) deviation? Probably the best way is to think of it in general terms as an "average," recognizing that, as I have previously pointed out, different averages give somewhat different numerical results. In general, the SD will be a little larger than would be expected for an average deviation, particularly in distributions that have many extreme scores.

This is because the squaring process, being multiplicative, tends to place more weight on extreme values than do additive processes such as the Arithmetic mean. For example, the AD for the distribution above is $190/80 = 2.38$, as compared to the SD of 2.89. However, keeping in mind that the SD is simply a form of average deviation of the individual values from their own mean is still the best way to develop a feeling for the size of this measure. As before, of course, the larger the SD is, the more variability there is in the distribution, and conversely, the smaller it is, the more homogeneous the scores are.

I want to make one final comment about the SD, and one that refers to those interesting properties I mentioned some pages ago. The square of the SD turns out to be the "second moment" of the distribution. [You will recall that we spent some time discussing the (first) moment nature of the arithmetic mean in the previous chapter. It is not necessary for us to delve into the specifics of this fact here. My purpose in noting it is simply to inform you that the mean (arithmetic) and the SD are natural companion measures with a common theoretical/mathematical basis.

For this reason, they are usually reported together, rather than being mixed in with other possible measures. Thus, it is common to hear of someone reporting the "mean and standard deviation" of a distribution when attempting to convey information about the central tendency and variability of that distribution.

B. Frequency-Based Measures of Variability

Just as the mean and standard deviation are members of the moment system of statistics, the median too has companion measures of variability in the frequency-based system of statistics. In order to discuss variability in the context of the "frequency system" of statistics, we will need to make a rather extensive digression to develop the idea of "percentiles."

1. Percentiles and percentages

As you might suspect, "percentiles" depend on the idea of percentages. Since most of us understand what a percentage is, you should remember that this topic is really quite easy, and rooted in considerable everyday experience. Just to be sure, however, percentage comes from the Latin for 100 (century—100 years, and cent—1/100 of a dollar, come from the same root). Percentage can be translated as "per 100," or "by the 100." When we take a percentage, we implicitly divide whatever it is into 100 parts, and then state how many of those parts we are talking about.

As an example, suppose we think of a pie. If I say that I'll let you have 75 percent of the pie, that means literally, 75 100ths of the pie. Of course, we don't actually have to cut the pie into 100 pieces and count out 75 of them in order to do this. But, theoretically, that's what 75 percent (75%) really means.

Percentage is simply a way of thinking about things when we want to talk about some portion of a whole. It is useful to remember that the object of the preposition "of" is the thing that is to be conceptualized as 100 parts, or the "whole" thing. Keeping this in mind avoids possible confusions about just what is the portion and what is the whole. For example, we say 75% of the pie, or, "10% of the labor force," or, 50% of the scores" and by this we mean that the whole is, respectively, the pie; (the number of people in) the labor force; and, the (number of) scores (in some defined data set).

Percentages are very useful because most people understand what you mean when you state a percentage, and, also, because dividing a whole into 100 parts permits, for most purposes, sufficiently fine distinctions that we can use whole numbers instead of fractions in making our statements. (At most, we use 10ths of a percent as the finest distinction made.) For example, we could mathematically talk about 13 1/4 25ths of the dollars in a budget being allocated to food; but, it's easier and more understandable to say 53%, the equivalent percentage.

2. Percentiles

So, what is a "percentile?" Let us assume that we have some kind of dimension that is represented by a score scale, ranging as usual, from low scores to high scores. Then a percentile is "a value on that scale such that a given percentage of the scores in that data set are smaller than that value."

To state it more formally: *The nth percentile* (P_n) *is the score value BELOW which n% of the scores in the distribution lie.*

Note that "*n*" can take any value *between* 0 and 100. The 0th percentile and the 100th percentile are indeterminate and meaningless, since *any* possible score below the smallest score in the distribution would have *no* scores below it, and *any* possible score above the highest score in the distribution would have *all* of the scores below it!)

Percentiles are members of the "frequency" system, since the actual values of the scores do not enter into the computation, merely their positions in the distribution—even though in computation we frequently estimate the exact values corresponding to the percentiles of interest.

Table 5-2. Percentile Data

Score Value	f	cf	Deviation (d)	d^2	fd	fd^2
50	3	80	5	25	15	75
49	8	77	4	16	32	128
48	6	69	3	9	18	54
47	10	63	2	4	20	40
46	10 37	53	1	1	10	10
45	12 12	43	0	0	0 +95	0
44	7	31	−1	1	−7	7
43	6	24	−2	4	−12	24
42	4	18	−3	9	−12	36
41	6	14	−4	16	−24	96
40	8 31	8	−5	25	−40 −95	200
N (Total)	80				0	670

By now, all of this should sound a little bit familiar—because we have encountered a percentile once before, in the previous chapter. Think (or look) back to the median. We defined the median as a point on the score scale where half (or 50%) of the cases lay below the point. But, this would be just the definition of the 50th percentile (P_{50}), wouldn't it? So now we have the obvious answer as to why measures of variability based on percentiles have the median as their natural complement. Indeed, the median and the 50th percentile are one and the same.

Before we go ahead to consider how percentiles are used in connection with estimating variability, we had better review how we computed a median, and extend this thinking to other percentiles. We'll use the same distribution we used earlier, and I present it again in Table 5-2.

Let's do three percentiles from this distribution, to illustrate the technique. Suppose we choose P_{35}, P_{50}, and P_{90}. Thinking about what a percentile is, and considering P_{35} for now, you should recognize that what is needed is a point on the score scale such that *35% of the cases lie below that point*.

Another admonition that we have made several times before is that whenever we do these computations on a distribution of scores, we have to remember that the discreet score values generally represent a continuous, underlying dimension. This means that a score of 46, say, must be thought of as a range of 45.5–46.5 on that dimension.

(Only if you do this can you have a continuum, since by defining each score this way, each point begins where the previous point left off, and ends where the next one begins. For example 46, 47, and 48 become 45.5–46.5, 46.5–47.5, 47.5–48.5. Only in this way can you avoid gaps between the numbers. You'll see why this is important when you begin to actually calculate a percentile.)

Now, back to P_{35}. We have a total of 80 scores, and we want to find a point where 35% of them lie below. 80 times .35 = 28, so where is the point on the distribution such that exactly 28 scores lie below that point? Counting up from the bottom, the scores 40–43 give us: $8 + 6 + 4 + 6 = 24$, leaving us 4 short of the 28 we want; but, if we take the next score, 44, we get an additional 7, or 31, which is too much. Thus, it is clear that the exact point where 28 scores lie below is buried in the seven 44 scores. What we are going to have to do is *interpolate*.

But, how do you interpolate within a single score? Now you know one of the reasons that we insisted on defining a single score as a range on the dimension rather than a point! Since 44 is defined as 43.5–44.5, it becomes possible to use the interpolation technique to estimate the exact place within this one point range where 28 scores lie below it.

If we have 24 scores up through 43 (really 43.5), and we need 4 more to get to 28, how many are available in the target range, 44 (really 43.5–44.5)? The table shows 7, so we need 4 of the 7, or 4/7 of that range. (Notice that in this sort of interpolation, the proportion is always formed from the ratio of the additional frequencies needed to those available in the next score range.)

Now all that remains is to apply the proportion to the score range we are working with: 4/7 times 1 = .57, and add this piece from the target interval to the lower limit of the interval within which you are interpolating: .57 + 43.5 = 44.07, which, then, is P_{35}. (Look at the illustration given below. You might also find it instructive to turn back to the preceding chapter and review the computation of the median. You will find it entirely analogous—the only difference being that in that case we started with 50% of the scores as our requirement rather than 35%, as here.)

Although we have done this earlier during the discussion of interpolation, perhaps it is useful to look at this problem laid out horizontally along a dimension line. Below are the data illustrating the problem of finding P_{35}.

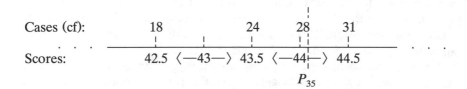

In other words, we had to go up through the distribution to 43.5 to get 24 cases; we needed 4 more of the 7 available at the next score interval; since we needed 4/7 of those cases, we also needed 4/7 of the score range in the next interval, or .57 of that score range (here, 1); thus, the required point on the scale is represented by the beginning of the score interval, plus the estimated 4/7 of it, or 43.5 + .57.

(Since our distribution was arrayed in terms of single score points, each of them represents an interval of one; if we had been using grouped data, the score range would have been whatever grouping interval we had used in the distribution. For example, if we had grouped our data by intervals of 3, then the score range in the target interval is obviously 3; if by 5, then it would be 5, etc.)

I might note in passing that if you have a number of these percentile calculations to do, they are facilitated by adding a column to your table of data, the cumulative frequency column. Cumulating the frequencies up from the bottom means that you don't have to count up the whole distribution each time you want to determine a percentile. This is why I included a *"cf"* column in Table 5-2.

Now, using the data in Table 5-2, we'll take a look at the median (P_{50}). Going through the same process, 50% of the cases is .5 times 80, or 40. Looking at the table, we can see that up through score 44 (upper limit, 44.5) we have 31 of the necessary 40, and through 45 (upper limit, 45.5) we have too many, 43 cases. Therefore we are going to need 9 (40−31) of the 12 cases available in the 45 score range (44.5–45.5). Well, 9/12 is .75 of the 1-point score range; so, $P_{50} = 44.5 + .75 = 45.25$.

In passing, you may recall that I set this distribution up with a mean of 45. Notice that the median is slightly larger, 45.25. The difference between the two tells us that the distribution is not quite symmetrical; and the fact that the mean is smaller than the median tells us that there are somewhat fewer but more extreme scores in the lower part of the distribution, which implies something of a negative skew (the distribution is somewhat more stretched out on the lower side).

With regard to this latter point, the half of the scores below the median to a certain degree "weigh" more than the half above, because some of them are somewhat extreme, and this pulls the mean down below the median a little bit. You can see from the table it requires 37 scores on the upper end to balance the lower 31 scores (the 12 45s don't add any weight to either side).

To complete the three examples we set out to look at, let us now find P_{90}. Again, we multiply the required percentage (90) times N; this gives 72 cases required. Running up the *cf* column, we have to go through score 48 to get 69, so that we need 3 of the 8 available in the 49 score range. Thus we have 3/8 of a score range of 1, which is .375. Adding the .375 to the beginning of the 49 interval, we have .375 + 48.5 = 48.875, rounded to 48.88 as our value for P_{90}.

3. Variability measures based on percentiles

It should be clear why a discussion of percentiles is appropriate in a chapter on variability. The variability measures in the frequency-based system, which are so companionable to the median, are indeed based on percentiles—*in fact they are percentile ranges.*

The range was the first variability measure that we discussed above. It was defined as the difference between the highest and lowest scores. Another way to state it would be to say the range is the number of points of score (highest minus lowest) that it takes to encompass 100% of the distribution (all of the scores). We talked about its weakness due to extreme scores. Also, we alluded to a "modified" range formed by discarding "outlier" data points, and our dissatisfaction with such a procedure.

However, if we could define some acceptable, more or less standard, modified range, rather than just arbitrary, ad hoc discarding of outliers, it might be a usable measure. Percentile ranges can be used for this purpose, and two percentile ranges have come into common use: the "P_{90}–P_{10} range" and the "average quartile" or "semi-interquartile range." Let's look at the former first.

a. P_{90}–P_{10} range. We know that 10% of the cases lie below the score corresponding to P_{10}, and that 90% of the cases lie below the score corresponding to P_{90}. So, this range is the number of score points it takes to capture the middle 80% of the distribution.

How is this a measure of variability? If this range is narrow (small), it means that 80% of the scores are crammed into a short space on the scale, or dimension, and thus are greatly bunched up toward the middle. If it is large, then the scores must be spread out a great deal more. Because it is the middle 80% of the cases, it is not unduly affected by a few outlier scores that might occur above P_{90} or below P_{10}.

b. The average quartile (semi-interquartile range). As you may have guessed, quartiles are the score points that divide the distribution into quarters, and of course are the same as the 25th percentile points: $P_{25} = Q_1$; $P_{50} = Q_2$ (or the median); and $P_{75} = Q_3$. The fourth quartile is left undefined, for the same reasons that P_{100}, which it theoretically would equal, is, that is, *any* score above the highest score would fit the definition, making it ambiguous. (The same argument applies to Q_0, or P_0.)

Now, an average quartile would be the number of score points necessary to encompass an average 25% of the distribution. Of course we don't have many quarters available to average, since we can't define either Q_0 or Q_4, but we can get two: $Q_2 - Q_1$ and $Q_3 - Q_2$.

Then, if we add these two values and take the mean by dividing by 2, we have "Q," the average quartile. This is an estimate of the number of points of score necessary to capture 25% of the distribution. If it is large, the distribution is spread out, and if it is small, the distribution is bunched up.

Of course the semi-interquartile range is exactly the same thing. It says that if you take the range (difference) between the quartiles (here, $Q_3 - Q_1$), and then take half of it (semi of course means half), you get this measure. It is easy to see that this procedure is mathematically equal to the one used to get the average quartile, and of course the interpretation is the same.

Generally, these two percentile ranges are the ones usually used in conjunction with the median, as they stem from the same (percentile) system. They are the most common ones, but you can easily see that almost any percentile range you could make an argument for could be constructed and used similarly. One that is sometimes used, for example is $P_{93}-P_{07}$.

Mostly these ranges are symmetrical around the median, which is a logical consequence of the desire to avoid the effect of extreme scores, as well as the fact that most of the distributions that we work with tend to be relatively symmetrical in nature.

You may ask if percentiles are good for anything other than variability estimates, and who uses them. Good, I'm glad you asked that!

4. Uses of percentiles: Norms

Because they are based on the notion of percentages, which has a high degree of common understanding and acceptance, percentiles are widely used to report the results of various sorts of measurements. One of the most common applications in education is the use of percentile "norms" (reference tables of performance) to report the academic performance of students on various achievement tests and other sorts of measures.

We won't dwell on norms in detail, but I will tell you their basic function, and how percentiles fit in. Suppose your child were to get a score of 45 on a test of verbal reasoning. How do you know what this means? Is that a good score or a bad one? High or low? Obviously, 45 by itself is meaningless; you need some reference information—a basis for comparison! This is what norms do—provide a basis for comparison.

Generally, to provide a basis for comparison, a norm consists of the reported performance of some reference group of people on that particular measure in which you are interested. Clearly, however, two conditions must be met in order for a norm to be useful to you: First, the norm performance reported must actually be typical of the group that it is supposed to characterize. For example, if the norm performance were purported to be typical of carpenters, but those who actually took the test were mostly laborers, the norm would be misleading.

Second, the norm must represent a meaningful comparison for the person using it. For example, if you were a high school math student, you wouldn't be interested in comparing yourself with a norm based on college professors; such a comparison can be made, but generally a meaningful comparison is one made against a group of which either one is a member, or expects to become a member.

Norms may be presented in terms of any one or more of a number of statistics, but percentiles are a favorite. Go back to the case of the child who scored a 45 on the verbal test. If we have published data (norms) for similar children on this test, they provide a means for comparing this child's score to the performance of the group of similar children.

It works this way. The publisher presents a table in which he shows what percent of the children in the reference group fell *at or below* each of the possible test scores. Now, we can look up 45 in this table, and if it shows a norm value of 74, we know that our child performed *as well or better than 74% of the reference group of similar children*. Thus, this comparison lends meaning to the score of 45, and we know that it is a rather good score as compared to the performance of similar children.

A related term is often used—"percentile rank." This is the other side of the coin, used to describe a given score value. In the above example we would say that the score of 45 has a percentile rank of 74. The basic meaning is the same. If the score obtained by the child is such that it exceeds a given percentage of the norm group, say 63%, the child's score has a percentile rank of 63.

To summarize, a percentile norm tells you what percentage of the group studied performed at or below the indicated score, so that when you compare an individual's performance to the norm, you can tell what percent of the reference group the individual exceeded.

Using a norm based on the right reference group is exceedingly important, as you might imagine. I can give you a personal example. Years ago when I was in graduate school I took a test of clerical ability. My score was such that I performed at the 52nd percentile when compared to a random sample of people in general—just about as average as you can get! However, that same score, when compared to norms based on *clerical employees*, came our at the 4th percentile (I exceeded only 4% of employed clerks!). I quickly drew the conclusion that I should look elsewhere than clerical duties for my vocation.

C. The Concept of Variance: Some Preliminary Considerations

There is one more concept to introduce in this chapter, that of "variance." This is another statistical term with a precise meaning, as opposed to the term variability, which is somewhat more general. I want merely to introduce the idea here. We will pursue it in more depth in the context of relationships among variables, discussed in the next chapter. I introduce it here because, as its name implies, it is related to the subject of the present chapter, variability.

In this chapter we have dwelled on ways to describe a typical, representative deviation. None of these measures embodies the idea of total amount. All focus on a result that approximates a single value from the distribution (albeit a typical one), rather than an index of the total amount of measurement taken over *all* of the observations in the distribution. It is sometimes useful to have such a measure.

Simply adding up the scores does not provide an estimate of total variability. Neither does simply adding up the deviations for each score because in doing so you are dealing with original score scale units. Our best measure of variability, the SD, is based upon a multiplicative relationship (squaring). Although it isn't entirely clear at first what an index of the total variability ought to be, intuitively it would seem that it should retain the multiplicative relationship that characterizes the SD.

With regard to our preferred measure of variability, the SD, having to take the square root as the last step of the computation is the part that takes you out of the multiplicative relationship. Thus, if instead we consider the *next to the last step* in the computation of the SD, we then have the arithmetic mean of the squared deviations. It is this mean squared deviation, that is used as an index of the total amount of variability in the set of scores in the distribution at hand; we call it by the statistical term, "variance."

Note that, because the only difference between the variance (V) and the SD is the omission of the last, square-root step, it follows that $SD = \sqrt{V}$, and $V - (SD)^2$.

Setting aside for the moment the numerical aspects of the variance, how should one think of it? Fundamentally, the concept refers to the total tendency (as opposed to typical value) of the items or observations in the set being studied to differ from one to another.

Why is this of interest? In general, we are always interested in cause and effect. While it is usually exceedingly difficult to tease out precise cause-and-effect relationships, it is obvious that if there is no effect, there can be no cause/effect relationship. But, what is the nature of an effect? Simply put, an effect is identified as a change, or variation. *The statistical reflection of an effect is variance.* (It is true, however, that not all variance is attributable to cause/effect relationships—some of it can be a reflection of chance fluctuations and errors of measurement.)

Thus, it must be understood clearly that the undivided variance present in any set of measures is always the mixed result of many causes. As a consequence, the crux of research and more particularly of statistical analysis is very often to attempt to ascertain how much of the variance present can be ascribed to specific causes: how much of it is due to some factor of analytic interest; how much is due to other particular, known causes; how much to unknown causes; and how much is attributable to pure chance or to errors of various kinds.

In fact, a branch of statistics has come to be known by the term analysis of variance, and focuses on the effort to partition the variance into a set of measures in ways that will at least reveal relationships, if not causality.

We will examine this subject in more detail in future chapters.

III. Summary

In this chapter we have seen the third of our basic statistical procedures for eliciting information from a basket of numbers, measures of variation. Measures of variability are of interest to us in their own right—quality control issues are a good example. Otherwise, they serve to inform our efforts to assess central tendency in a distribution by aiding us to decide how typical a measure of central tendency really is (or, to what extent central tendency exists in that distribution).

We looked at several measures. The range is simple but imprecise—heavily affected by even one extreme observation. Applications of the averaging concept led us to the average deviation, which is theoretically weak because of its peculiar treatment of negative deviations. We found that by using the quadratic mean deviation the problem with negatives could be satisfactorily overcome, yielding a measure that has become known as the standard deviation—a member of the same mathematical system (moments) as the arithmetic mean.

There are companion measures for the median as well. The chapter points out that these measures are based on the percentile, or frequency, system. In this system, the scores are the first arrayed, and then score points are determined below which specified percentages of the cases in the distribution lie. These points are named the nth percentiles where n is the percentage of cases lying below the point. As variability measures, interpercentile ranges were examined, which yield a range of score within which a particular percentage of the cases is to be found. $P_{90}-P_{10}$ and the semi-interquartile range $[(P_{75}-P_{25})/2]$ are the most popular.

Finally, the concept of total variation, or variance was introduced. The measure is simply the square of the SD, but serves as the basis for more sophisticated analyses to be described in later chapters.

IV. References

Johnson, Robert R. *Elementary Statistics*, 5th Ed. Boston: PWS-Kent, 1988.

Levin, Jack. *Elementary Statistics in Social Research*, 3rd Ed. New York: Harper & Row, 1983.

See also References, Chapters Three and Four.

III

Covariance, Relationships, and Inference

In Part I we discussed the fundamentals underlying statistics, numbers and how to manipulate them, and data and where they come from. In Part II we dealt with the basic statistical procedures designed to illuminate the information contained in a set of numbers, or data, derived from some source of interest. Here we considered distributions of various kinds, measures of central tendency, and measures of dispersion or variability.

In Part III we are going to move on to the consideration of more complex (and more real life) problems and situations. Chapter 6 begins with a discussion of co-variation, the tendency of two variables to vary together in some related way over a set of data. It is, of course, common experience to note that some things seem related, such as height and weight, or intelligence and success. The chapter continues with a discussion of how to quantify and measure the extent of such relationships. It concludes with a discussion of how we might employ such relationships to predict future events such as success in college or business, etc.

The second chapter in this Part, Chapter 7, extends the discussion of relationships and prediction. It begins with a brief discussion of some of the less common indices of relationship and their applications. Then we move to a discussion of a very common statistical technique, contingency analysis, or the use of the Chi-square statistic. Following this we look at the fascinating topic of relationships among more than two variables. Finally, we finish up with a discussion of some reflections on the interpretation and limitations of the correlational process.

Chapter 8 concludes this Part. It is devoted to one of the most pervasive statistical topics. Here we consider statistical inference, which, in general, involves the estimation of one or more traits of a population when only a subset (sample) of that population has been actually measured. Aspects of estimating the error in such population projections are also examined.

The material in Part III is not independent of that covered in Part II, since most of the concepts and measures covered in Part II become the subject of, or are involved in, the more complex procedures covered in Part III. The material here is somewhat more difficult because of its increased complexity, and the presence of some concepts which may be unfamiliar. Nonetheless, patience will be rewarded. The ideas covered in these chapters are surely of paramount importance to those who wish to understand the role of statistics in today's world.

A final admonition before beginning. You will have noted that I have tried to simplify things a little by keeping the notation as simple as possible. I have used "SD" to stand for standard deviation (instead of the more widely used σ) because I think it is easier to remember the abbreviation than the Greek. Similarly, I will use "V" to stand for the variance in the following material instead of the more common σ^2. Please understand that there are many specialized and alternative notations, and covering them all could interfere with our main task of developing an understanding of the key concepts we are discussing.

6

Covariation and Correlation

I. Overview and Purpose

Up until now, we have been concerned with a descriptive process. We have developed several ideas pertaining to the problem of describing "a basketful of numbers." The three major suggestions were: to examine distributional considerations; to estimate the central tendency; and to look at the variability. We have reviewed a number of aspects of each of these three major areas.

Now, however, it is time to branch out some. The present chapter retains the emphasis on *describing* the information contained in a set of numbers, but looks beyond the examination of single variables one at a time to inquire into the description of the *relationship* among two variables jointly. Of course, each of the single variables involved in such relationships can still be examined in the ways we have discussed in the last few chapters. But, what we are after now is a way, or ways, to describe the relationship between two variables jointly—in a sense to examine their interactions with each other.

Why should we be interested in such a topic? Because it is here that the most fascinating aspect of statistical research and investigation resides. When we study relationships, we have the possibility of discovering that knowledge about something we know well will be found to be linked to knowledge about another perhaps less well-known thing. Thus, we have the possibility of developing a web of interlinked knowledge. Ultimately, we have the possibility of discovering and understanding to some degree what causes what. Only with the knowledge of relationships can we develop theories, understandings, and the potential of controlling and shaping events to such ends as we may desire.

In addition to developing the ideas involved in joint, co-relationships, we will, of course, be interested in constructing a numerical (statistical) index to measure that relationship. Also, we will consider how to use a relationship to predict unknowns. Finally, we will look at some of the limitations that exist in talking about the relationships among statistical variables.

The topics that we shall examine in this chapter are several, and are closely related, so that it is somewhat difficult to know where to start. So, like a swimmer jumping into a lake, we shall just dive in, and hope that we get around to all of the parts of the lake in good order. Some patience will be required.

II. Variance and Covariance

We ended the last chapter with a preliminary discussion of the tendency of scores to vary or differ from each other (variance). The major thrust of the first portion of this chapter is to establish firmly the importance of the variance concept with respect to the interrelationship among variables, called "co-variance." "Co" means together, and thus "co-variance" means "varying together," and we must examine the extent to which two sets of scores exhibit similar tendencies to differ among themselves (co-vary).

(For now, we will confine ourselves to looking at sets of two variables at a time. To be more specific, we want to look at the situation where you have *one* group of items or individuals, for each of which you have a score on *two* different dimensions or variables. Such sets of paired scores are called "bi-variate" or "joint" distributions.)

First, let's see that variance and covariance are similar in a numerical respect. Suppose we have four variables; call them A, B, C, D. We want to examine the relationships among them, but we have agreed to take them two at a time. Therefore, we want to look at the relationship between A and B, A and C, A and D, B and C, B and D, and C and D. (These are always commutative relationships; that is, the relationship between A and B is numerically identical to that between B and A, and so forth.)

I could put this into a square display with the four variables listed across the top of the display and again, in the same order, down the side of the display. Such a display is called a "matrix," and the entries in the body of the display, called "cells," contain all of the various pairwise combinations of the four variables. The entries running from the upper left corner to the bottom right corner are called the "diagonal cells," or just "the diagonal."

Of course, the bottom half of the matrix (below the diagonal) really duplicates the information contained in the top half, and could be (and usually is) omitted. [The reason for the duplication is the commutative property ($AB = BA$, etc.); if AB were *not* equal to BA, then we would have to have the complete matrix in order to cover all of the pairwise possibilities.] Look at the display as follows:

	A	B	C	D
A	AA	AB	AC	AD
B	$\underline{BA}$	BB	BC	BD
C	$\underline{CA}$	$\underline{CB}$	CC	CD
D	$\underline{DA}$	$\underline{DB}$	$\underline{DC}$	DD

Notice that the entries in the matrix are comprised of the row headings taken with the column headings, pair by pair. All of the pairwise relationships among these four variables, A, B, C, D, are displayed in this matrix. (The underlined cells are the ones that are duplicative with their counterparts above the diagonal.)

The entries on the diagonal are of interest to us. Although each of the other cells in the matrix shows a combination of two variables (and is thus a "co-variance" cell), the diagonal is comprised of each variable taken with itself. *These are the variance cells.* For example, *"BB"* stands for the variance of the variable *"B,"* computed as we discussed in the previous chapter, etc.

The idea that I want to convey here is simply that variance and covariance are intimately related concepts. In fact, the variance cells are really a subset of the covariance pairs: they are each a variable in the set, paired with itself. They are always found on the diagonal of *any* such matrix.

I have introduced this little matrix to bring home a very important idea (which I will explain in the course of this discussion). The covariance cells look like the multiplication of two variables, for example, *ab*, *cd* or *xy*, whereas the variance terms look like the multiplication of one variable times itself (squaring), for example, *aa*, *dd*, or *xx*, or *yy*.

(In passing I should note that mathematical statisticians would undoubtedly employ much more sophisticated techniques to achieve the development of covariance and its related index than I intend to employ in the next few paragraphs. However, I intend to develop this matter from an arithmetic viewpoint, since the average reader is more likely to understand that process than the application of advanced mathematics.)

(It is for this reason that I emphasize the multiplicative nature of covariance in the treatment to follow. I believe that the multiplicative nature of these relationships offers a valuable way to look at both variance and covariance, and we will spend some time developing the argument that it originally arises in response to those pesky little minus signs that forced us away from the average deviation and to the standard deviation.)

A. Variance from Scores

When we developed the SD, we were looking for an *average deviation* from the arithmetic mean. We were forced into using squared deviations in order to deal with minus deviations *in a mathematically more acceptable way*. At that time I did not trouble you with the fact that *dealing with squared deviations is arithmetically equivalent to dealing with squared scores in calculating a variance*. Now, however, because I want to talk about scores, I will demonstrate the equivalence in the box below.

Suppose we deal with, as usual, a variable we'll call "X". A deviation (d) from the arithmetic mean (call it "M") of the Xs could be written as follows:

$$d_i = (X_i - M_X),$$

where the "i" simply refers to any one individual in the distribution. So, for any particular individual, his deviation is his own X-score minus the mean of the Xs. Following along, his squared deviation would be:

$$d_i^2 = (X_i - M_X)(X_i - M_X).$$

Multiplying this out gives:

$$(X_i)^2 - 2(X_i)(M_X) + (M_X)^2.$$

Now, we consider a set of such scores. To get Σd^2, as we would have to do in calculating the SD, we sum this expression over all of the individuals in the data set. We would get for the first term, ΣX^2; the second term becomes, $(-2M\Sigma X)$; and the third term becomes, $+NM^2$. These three terms are the expanded form of our previous form, Σd^2:

$$\Sigma d^2 = \Sigma X^2 - 2M\Sigma X + NM^2.$$

However, remember that the mean is $\Sigma X / N$, and if we now substitute this value in the second and third terms, we have:

$$\Sigma d^2 = \Sigma X^2 - 2\Sigma X \Sigma X / N + N \Sigma X \Sigma X / N^2.$$

The latter two terms can be combined if we divide both the numerator and denominator of the third term by N, so that they have a common denominator of N: $(-2\Sigma X \Sigma X) / N + \Sigma X \Sigma X / N = -\Sigma X \Sigma X / N$. This gives us the final result of expanding:

$$\Sigma d^2 = \Sigma X^2 - \Sigma X \Sigma X / N.$$

To get the variance, however, remember that the sum of the squared deviations must still be divided by N (or $N - 1$, if you want to be mathematically precise). Doing this to the above expression yields:

$$\Sigma d^2 / N = V = \Sigma X^2 / N - \left(\Sigma X / N \right)^2.$$

To translate this finding into words, consider this. The first term is the arithmetic mean of the product of the scores times themselves, that is, the squared scores (rather than the mean of the squared deviations that we dealt with in the previous chapter). The second term is the square of the arithmetic mean of the scores.

It will be recalled that the mean was used as the reference point on the scale for the calculation of the deviations we originally worked with. By squaring it, we translate it into the same units (squared score values) as those characterizing the squared scores. Thus, as defined from the score values rather than the deviations, the variance retains both the aspect of squaring and its relative relationship to the mean of the distribution as the reference point.

Although all of this may be somewhat obscure, the point is that *the essential information comprising the variance (whether you start with scores or deviations) arises from the squares of the scores — a multiplicative relationship in which this variable, X, was taken times itself.*

B. Multiplication and Relationship

Multiplication (squaring) arose in my development of an index of the variance (as a valid way of handling negative deviations). Although other, more complex indices might be developed, the variance as presently defined appears to be the simplest, mathematically tractable approach to the problem of trying to measure the tendency of scores to differ from one another.

When we deal with two variables measured on the same population at the same time, each set of scores, of course, has its own tendency to be different from each other (vary). When we ask about *co*variance, we are asking how similar those two tendencies to vary are. In measuring this, it would be convenient if the measure of covariance were complementary with the measures of variance we use to describe the two sets of scores individually.

To have an index that is complementary to the variance, we can define it in a similar way, using a multiplicative relationship. This is true even though we do not have to be concerned about whether or not there are negative signs in the system.

Thus, a multiplicative relationship was suggested by the variance, because of the negative deviations; and, a multiplicative relationship is suggested by the covariance, because of its similarity to the variance, even though negative signs are not at issue. Now, let's see how the use of multiplication provides a reasonable procedure for these problems.

1. Multiplication and the variance

Considering first the variance (V), it should be true that if all of the scores are the same (no tendency to differ), a numerical index of the variance should result in zero. As defined in the previous chapter, it does. For example, look at the following little distribution:

Score	Squared Score
6	36
6	36
6	36
6	36
24	144

The mean is: $\Sigma x/N = 24/4 = 6$.
V is: $\Sigma x^2/N - M^2 = 144/4 - 6^2$, or $36 - 36 = 0$!
That's as small as you can get!

In addition, the index we use should provide the maximum numerical value of any of the possible multiplicative arrangements of the scores, so that we can be sure of capturing the full variability in the set of values. Thus, we should get the maximum value for the index when each score is squared, as opposed to being multiplied times one of the other scores. Look at the following sets of multiplications:

Score	Times	Equals	Times	Equals	Times	Equals	Times	Equals
7	7	49	6	42	7	49	7	49
6	6	36	7	42	6	36	4	24
4	4	16	4	16	3	12	6	24
3	3	9	3	9	4	12	3	9
20		110		109		109		106

The first arrangement mean is $20/4 = 5$.
The first arrangement V is $\Sigma x^2/N - M^2$; or, $110/4 - 25 = 2.5$.

Here I have taken the values in the column labeled "Score" and multiplied them by the values in the various columns labeled "Times," where these latter values are rearrangements of the original four scores. (I have not included all possibilities here, only some to illustrate the point. Feel free to try others, if you wish.)

In each case the sum of the products is less than that obtained from the calculation of V given below the distribution. You can see from this illustration that no arrangement of multiplying these four numbers gives a result as great as that obtained by squaring them. Therefore, no other arrangement of multiplying these scores could give a variance as large as squaring them does!

Thus, V, as defined, meets the test of reasonableness, in that it gives zero as a result where there is no tendency among the scores to differ, and gives the maximum possible result of all possible multiplicative pairings of the scores.

2. Covariance by multiplication

Now let's examine the problem of expressing a relationship between *two* variables, measured on the same population. This means that each individual or item in the population is represented by two scores, one for each of the two variables. Logically, if the two variables are closely related, such as height and weight or education and success, we would expect those scoring high on one to also score relatively high on the other; likewise, those scoring low on one would tend to score relatively low on the other. Also, if one of the variables should just be the other by a different name, we should expect that the variations among individuals in one set of scores would track those in the other set perfectly.

On the other hand, sometimes relationships are what we call "inverse." In such cases, there is a tendency for those doing well on one variable to do poorly on the other, and vice versa. A crude example might be number of days sick per month versus a measure of job success (the higher the first, the greater the tendency for the second to be lower).

A numerical index of covariance should reflect both the "direct" and the "inverse" types of relationships. What we want to show is that multiplication of these scores provides a way of measuring the extent of such relationships.

Suppose you won a prize, and for your prize you were allowed to fill three buckets with tokens and then to trade them in for the amount of money represented in your buckets. Suppose further that the three buckets are a 1-gallon, a 2-gallon, and a 5-gallon size. Suppose that the tokens, although of the same size, are worth $1 for whites, $5 for reds, and $10 for blues.

Intuitively it seems clear that one would fill the largest bucket with the most valuable tokens, and the next with the next, and so forth, in order to maximize the money won. Thus, the following arrangement would provide the most money:

Bucket Size	Token Value	Value Units
5	10	50
2	5	10
1	1	1
		61

Although we don't know exactly how much we've won until we're told how many tokens there are to a gallon, there is little doubt that we have maximized our prize, since no other arrangement of buckets vs. token value produces a total number of value units as large as 61. Try the other possible allocations of bucket size to token value and satisfy yourself that this is true. There are five other arrangements, and they produce totals of 57, 46, 37, 30, and 25.

Notice also that the minimum payout occurs when the numbers are exactly reversed; that is, 5×1, 2×5, and 1×10.

Now, suppose we take any two sets of numbers, call them x and y. The product of each x and its corresponding y represents the "amount" of the grand total attributable to that xy pair, just as the product of bucket size and token value represents the portion of the payout due to that pair in our example. Again, the maximum possible total of the xy products occurs when the largest x is paired with the largest y, the next largest with the next largest, and so on down to the smallest with the smallest.

Again, similarly to our example, the very smallest grand total occurs when we pair the xs and the ys inversely, the largest with the smallest, the next largest with the next smallest, etc.

There are of course ways to combine the two sets of numbers that would provide larger totals. You could, for example, square one distribution before multiplying it by the other. However, doing so would give the squared distribution much more of an effect in the final grand total. Intuitively, it would seem that this would be undesirable. Generally, there would be little or no rationale for weighting one or the other distribution differentially in assessing the relationship between the two.

Thus, it follows that we can employ this multiplicative relationship, specifically the sum of the xy pair products, to assess the extent to which the values of the two variables track each other from large to small. The larger the sum of the "cross-products" (xy pairs), the stronger the tendency for scores on one variable to be like those on the other.

3. What about negative values?

In dealing with the variance, we used the mean of the distribution as a reference point for the deviations we squared. Because the squaring eliminated all negative deviations, the variance is made up solely of positive cross-products (xx pairs, or the distribution cross-multiplied with itself). It follows directly that the *variance can be only a positive value*—no negatives ever enter into its composition. This makes intuitive sense as well, as it would be impossible for the score values to differ among themselves less than not at all!

Covariance cross-products (xy pairs) are a little different, however. (I have demonstrated that it makes no difference in working with the variance whether we use deviations or raw scores; it is possible to demonstrate that the same is true for covariance, but I ask you to take my word for it, in the interests of time and space.)

Suppose you think of the cross-products as being products of the deviations from the means of each variable instead of products of raw scores. Then an xy cross-product might be something like this: $(dx)(dy)$, or $(x - M_x)(y - M_y)$. Now, this would mean that the numbers being multiplied together might be either pluses or minuses, depending on whether the x and y scores were above or below their respective means. [The problem of adding together to zero doesn't exist here, because the xs are deviations from the x mean and the ys are deviations from the (different) y mean!]

This means that the *covariance terms may be either plus or minus*, even though the variance terms are restricted to only positive values. Indeed, there is a great advantage in this aspect of the covariance terms. Suppose that all of the scores above the mean on x were paired with all of the scores below the mean on y. This would be an inverse relationship (low scores on one variable going with high scores on the other, and vice versa), as we mentioned earlier.

However, the sum of the cross-products would tell us this! The scores above the mean would have positive deviations and those below would be negative. Thus the xy pairs would tend to be negatives. (We remember that a plus times a minus yields a negative product.) So, the sum of the cross-products in this instance would be negative, signaling the inverse nature of the relationship.

Thus, we have seen that the multiplicative relationship is sensitive to the position of scores in one distribution relative to the position of the other member of the score pair in another distribution. Finally, this relationship is a cardinal function, not just a matter of order in the distribution, so that the cross-product total assesses not only the extent to which scores are in the same order in the two distributions, but the extent to which they are proportionately in relatively the same positions in those distributions.

To close this part of our discussion, we return once again to the ubiquity of the average in statistical developments. How do you suppose the concept of cross-product sums is turned into an index of relationship? Yes, after adjustments for differences in means and units on the two variables (discussed later in this chapter), the index of relationship most widely used simply boils down to the arithmetic mean of the cross-products; it is also known as the "product-moment correlation coefficient," symbol: "r".

The adjustments referred to also have the effect of limiting the total sum of the cross-products to an amount equal to the N of the joint distribution. Therefore, the maximum value for "r" is equal to N/N, or $+1.0$, which corresponds to a perfect, one for one, relationship between the two variables (a situation that never occurs in real life unless one of the variables is simply the other in disguise).

The same adjustments also limit the minimum size of the total sum of the cross-products to a value equal to minus N, so that the minimum value of "r" is $-N/N$, or -1.0. This value indicates a perfect inverse relationship, which, again, never occurs naturally.

When the distribution of cross-products is essentially random, so that the subtotal of positive terms is balanced by the subtotal of negative terms and together they total to zero, then "r" is of course also zero ($0/N = 0$).

The zero condition indicates that there is no systematic relationship between the two variables, that the tendency to vary found in one variable is unrelated to or "independent" of that found in the other. Such variables may be thought of as "uncorrelated" or "independent" of each other. It means that knowing something about one tells you nothing at all about the other.

To sum up, then, both variance and covariance are measured by summing the products of two numbers for each of the individuals in a study; the former is based on squaring the scores in a single distribution; and the latter, on multiplying the scores on two variables for each individual.

In each case the sum is divided by N to get the arithmetic mean of the products (although, if raw scores are to be used, some adjustments must be made in the distributions in order to compensate for differing units and reference points). The possible numerical value of "r" ranges continuously anywhere from -1.0 through 0.0 to $+1.0$, or from perfect inverse relationship through no relationship to perfect direct relationship. We will return to the discussion of "r," approaching it from a different perspective, later in this chapter.

III. Some Distributional Relationships: Standard Scores

Because I have several times mentioned the matter of "adjustments" made to a distribution in going from deviation units to raw score units, and so forth, it is appropriate to present a brief discussion of certain distributional relationships that underlie such a process.

The reason for doing so here is not just because it is important in developing our notions about correlation, but because it also provides an explanation of the concept of "standard scores" that may be of interest in its own right. For example, educational tests such as the Scholastic Aptitude (SAT) tests (and other psychometric instruments as well) frequently use standard scores in reporting their test results.

The basic relationship that we will explore here is what happens to the statistics of a distribution, such as the arithmetic mean and the standard deviation, when (each score in) the distribution is subjected to an arithmetic operation (adding, subtracting, multiplying, or dividing) *by a constant* (the same number applied to each individual score).

Suppose we illustrate with a little distribution of five scores, applying a constant in various ways, and calculating the means, variances, and standard deviations:

Scores	Squares	Scores + 5	Squares	Scores × 3	Squares
4	16	9	81	12	144
2	4	7	49	6	36
7	49	12	144	21	441
6	36	11	121	18	324
1	1	6	36	3	9
Sum: 20	106	45	431	60	954

M:	4.0		9.0		12.0
V:	5.2		5.2		46.8
SD:	2.28		2.28		6.84

Now, let's see what we can learn from this display. First, look at the third column. In this column I have added the constant of 5 to each of the five individual scores, as shown. The mean of the original set is 4.0 and the mean of the altered set is 9.0. It is no coincidence that the mean of the second set is exactly 5 points higher than the original mean.

Let's take this apart. For each "x" I have added 5, call it $(x + 5)$. In calculating the new mean, I must sum up the five new values (using the rules for summation which we covered in Chapter 1):

$$\sum (x + 5) = \left(\sum x + 5N \right).$$

Then I divide this quantity by N to get the mean:

$$\left(\sum x + 5N \right)/N = \sum x/N + 5N/N.$$

This reduces to the old mean $(\sum x/N)$ plus the constant, 5, since the N/N cancels out.

However, if we look at the SDs for columns 1 and 3, we observe that they are exactly the same. This is intuitively reasonable, as the SD measures the tendency of the scores to differ from one another; and, although they have all changed, they have each changed by exactly the same amount, and thus their differences among themselves remain unchanged. We can verify this by again looking at what really happens.

Suppose we consider the variance (remember the SD is the square root of V), defined in terms of deviations: $\sum d^2/N$. Each deviation is defined as a score minus the mean of the distribution: $d = (x - M)$. Now, when we add the constant 5, a score would become $(x + 5)$, but we have just seen that the mean *also* becomes 5 larger, or $(M + 5)$. But then the deviation reduces to what it was for the original distribution:

old $d = (x - M)$

new $d = [(x + 5) - (M + 5)] = [x + 5 - M - 5] = [x - M]$.

Now, if each of the five deviations remains exactly the same, then the addition of the constant could not affect the SD.

Thus, we get a general rule:

Adding a constant to each score in a distribution *adds that same constant to the mean of the distribution*. (Because addition and subtraction are opposites, it can also be shown similarly that subtracting a constant from each score in a distribution *subtracts that same constant from the mean of the distribution*.) However, the standard deviation *remains the same* in both cases.

Looking at the fifth column, here I have multiplied each score in the original distribution by a constant of 3. The new mean is 12.0, and again it is no coincidence that this is exactly 3 times the old mean. Thinking back again to our rules for summation, each score is 3 times the old one or $3x$. When I sum these in the process of calculating the new mean, I have $\Sigma(3x)/N$ for the new mean. Thus:

$$\Sigma (3x)/N = 3\left(\Sigma x/N\right).$$

Since $\Sigma x/N$ is the old mean, the new mean is 3 times the old one!

Continuing to look at the fifth column, let's consider the SD. It may be observed that the SD of 6.84 is exactly 3 times that of the original distribution (2.28). Let's see if we can see how this happened.

Again consider the SD defined in terms of deviations. Multiplying each score by the constant of 3 multiplies the corresponding deviation by 3: if the original deviation is $(x - M)$, then the new deviation is $(3x - 3M)$ (since we have just seen that multiplying by a constant multiplies both each score *and* the mean by that constant). But, $(3x - 3M) = 3(x - M)$, or 3 times the old deviation (i.e., $3d$). We calculate the new V as follows:

$$V = \Sigma (3d)^2/N = \Sigma 9d^2/N = 9\left(\Sigma d^2\right)/N$$

$$SD = \sqrt{V} = \sqrt{9} \sqrt{V} = 3 \text{ times the old SD!}$$

Similar logic holds for the case of division by a constant.

All of this gives rise to another rule: When (each score in) a distribution is multiplied (*or divided*) by a constant, *both* the mean and the standard deviation are multiplied (*or divided*) by that same constant.

A. The Standard Distribution

Earlier I said that the correlation coefficient was, *with certain adjustments*, the arithmetic mean cross-product. Now we can state what those adjustments were. Each distribution must *first* be transformed into a "standard" distribution. A standard distribution is one that has a mean = 0, and a SD = 1.

The operations with constants that we have just gone over provide a means to convert *any* distribution into a standard distribution. Let's see how this works.

To transform any distribution to one in which the new mean is 0, you first compute the mean of the original distribution, and then you subtract that same value from each score in the distribution. We have just learned that subtracting a constant from each score in a distribution subtracts that constant from the mean. If the constant is equal to the mean, then the new mean = the old mean minus itself, or zero! And, according to our rules, above, the SD is unaffected.

Now we have a derivative distribution of a variable we might represent as $x - M$; it has a mean of 0, and the original SD. To transform this derivative distribution into one with a SD of 1, we simply divide each $x - M$ score by the original SD. According to our rules for constants, doing this will yield a distribution where the old SD will be divided by a constant equal to itself, so that the resulting SD will be $SD/SD = 1$! Note that division by a constant *also* divides the mean by the constant. *But*, because the mean is already equal to zero, $0/SD$ is still 0!

Thus, in two steps, we have converted *any* original distribution into one with a mean = 0, and a SD = 1. With respect to the original x-scores, we first subtracted the calculated original mean from each score $(x - M)$; then with respect to the distribution of $x - M$ values, we divided each one by the original SD, getting $(x - M)/SD$.

Case #	Original distribution	1st Transform	2nd Transform
1	y_1	$y_1 - 16$	$(y_1 - 16)/4$
2	y_2	$y_2 - 16$	$(y_2 - 16)/4$
$\vdots$	$\vdots$	$\vdots$	$\vdots$
N	y_N	$y_N - 16$	$(y_N - 16)/4$
Mean:	16	0	0
SD:	4	4	1

For example, suppose we have a distribution of "y," with a mean of 16, and a standard deviation of 4, to convert to a standard distribution. The display above breaks down the process.

To sum up, the distribution of $(x - M)/SD$ *always* has a mean of 0 and a standard deviation of 1, no matter what "x" stands for. *It is when both distributions are in this "standard" form that the mean cross-product is the correlation coefficient.*

B. Standard Scores

I mentioned that standard scores are often used by testing agencies to report the results of "standardized tests" such as the SATs used to assess students' prospects of success in college and various achievement tests used to assess student progress. When test scores are reported in "standard score" form, it means that the testing agency has converted the original test score distribution into one with predetermined statistics. There are several purposes for doing this. The most important one is that by standardizing the reference point (mean) and the units of score (SD) it becomes more meaningful to compare the scores of students on different test instruments than it would be to compare the raw scores.

(Note, however, that there are other issues that should be addressed in any attempt to compare test scores. For instance, one important issue is that to be truly comparable the two tests should have comparable content and purpose. Two tests can each be called "reading" tests, for example, yet contain widely different types of material. A second concern is that the two tests should have been given to the same or highly similar groups. Standard scores do not solve all of the potential problems!)

In fact, different tests are never more than approximately comparable, no matter what statistics are used. But, if the original distributions on two tests are transformed so that they have the same means and standard deviations, at least the comparison will be based on numerically the same reference point (the "standardized" mean) and numerically the same units (the "standardized" standard deviation).

Thus, you can ask questions such as: How far above (or below) the average score (mean) was this pupil's score in terms of multiples or fractions of the average score difference (SD) for all the pupils who took this test? For instance, if the standard score mean were 60, and the standard score SD were 15, and a student scored 50, you would know that he performed 2/3 of the average deviation of the comparison group *below* their mean performance level.

To make such transformations, the testing agency first converts to standard distributions just as we did above. Then they take two additional steps: first, they multiply the distribution by a constant equal to the standard deviation they wish to end up with. [This multiplies the old SD (1 at this point) by the constant, but does not affect the mean (0), as anything times zero is still zero.]

Second, they then add a constant equal to the mean they wish to end up with. [Thus, the new mean becomes 0 + the constant; but, of course, the addition operation does not affect the new SD.] Symbolically, the new standard score distribution, with whatever new mean and new standard deviation you have chosen for it, is related to the original x-score distribution as follows:

$$\text{New std. score} = [(\text{new SD})(x - M)/\text{SD}] + \text{new mean}.$$

To illustrate this process, consider the distribution that we set forth a couple of pages ago. There we started with a distribution with a mean of 16 and a SD of 4. Let me illustrate the two additional transforms necessary to give it a new mean and standard deviation of, say, 100 and 15, respectively:

Case #	"Standard" Dist.	Step 1	Step 2 ("Standard Score")
1	$(y_1 - 16)/4$	$15[(y_1 - 16)/4]$	$15[(y_1 - 16)/4] + 100$
2	$(y_2 - 16)/4$	$15[(y_2 - 16)/4]$	$15[(y_2 - 16)/4] + 100$
$\vdots$	$\vdots$	$\vdots$	$\vdots$
N	$(y_N - 16)/4$	$15[(y_N - 16)/4]$	$15[(y_N - 16)/4] + 100$
Mean:	0	0	100
SD:	1	15	15

Please note that although the above discussion is valid as far as it goes, it somewhat oversimplifies the actual procedures used by a testing agency. As an example, I had intended to discuss the procedures used by the Education Testing Service (ETS) when it develops a standard score distribution for the Scholastic Aptitude Tests (SATs). However, even though they do use the process we have discussed above, they do so many additional statistical adjustments and transformations that I cannot present the true situation without serious risk of being either overcomplicated or misleading.

Among these other procedures are adjustments intended to make the scores comparable from year to year and from subtest to subtest and from version to version, etc. Therefore, I will simply say that such conversions as above are among the procedures used by testing agencies in developing the standard scores that they publish along with their standardized tests.

C. Co-relationship from the Standpoint of Regression

The discussion in the first portion of this chapter was intended to give you a feeling for the arithmetic underpinnings of co-relationship, in terms of the underlying multiplicative processes and distributional relationships. There is at least one other major way to approach the subject of co-relationships, and, at least two other ways to develop the same index of correlational relationships ("r"). We will continue by examining the topic of "regression."

It is perhaps unfortunate that this area of our subject has been termed "regression," as it sounds vaguely undesirable. However, the term is well entrenched, dating back to the studies of the famous English scientist and theorist, Francis Galton. Galton was greatly impressed by the work of his cousin, Charles Darwin, whose theories on evolution were published in *The Origin of Species* in 1859. Setting himself to the study of hereditary factors as a result, he began to collect data on the stature of children versus that of their parents.

Galton soon found, not to his surprise, that tall parents tended to have tall offspring, as would be expected from considerations of heredity. However, rather unexpectedly, he also found that the offspring tended to be less extreme in height than their parents. In 1885, Galton published the paper in which the term "regression line" was introduced, "Regression toward Mediocrity in Hereditary Stature." Today we speak of "regression toward the mean" to describe the tendency of extreme cases to produce later generations that are extreme, but more like the population averages than their parents were. This principle applies across many fields.

(In passing, you may have wondered why the symbol "r" was introduced earlier in this chapter to stand for the index of relationship called the correlation coefficient. Although another early statistician, Karl Pearson, is far better known for his pioneering work with the correlation coefficient around the turn of this century, it was Galton who introduced the symbol r, which was derived from his hereditary studies and stood for the term "regression coefficient."

Incidentally, Galton is also credited by some with the introduction of the term "correlation" in connection with these same studies of hereditary stature. However, Galton's regression coefficient was not mathematically the same as the correlation coefficient in use today.)

Although Galton's application of regression was not exactly like the typical regression problem in today's statistics, it did involve the effort to establish a relationship between two variables, and, in effect, predict one from another on the basis of that relationship, using "regression lines." It is perhaps better for us to think about this area as "assessing predictive relationships," rather than to struggle with any semantic conflicts we may have regarding the term regression.

Hence, if it can be established that two variables are significantly related to each other, *knowing something about one enables you to make better-*

than-chance predictions about (or estimates of) the other—without independently measuring it. This is the crux of our interest as statisticians, scientists, or consumers of information!

The area of prediction or estimation can be approached by computation alone, but it is perhaps better understood if we approach it through the use of graphic devices. Come, let us see...

In previous chapters, we have seen that we can plot any distribution in a more or less standard fashion, by placing score categories on the baseline and displaying the frequencies in the distribution associated with each score category as tallies along the vertical axis. However, if we wish to study the relationship between two such distributions, we must accomplish this type of plotting *jointly, for both at the same time*.

This is accomplished by placing the score scale for one of the variables on the horizontal axis or baseline, just as before, with the smaller values to the left. However, the score scale for the second variable is then laid out along the vertical axis, with the smaller values toward the bottom. Then a point, or tally, for each of the score *pairs* is placed in the body of the plot at a location corresponding to the two score values.

The plot for such a joint, or "bi-variate," distribution is called a "scatterplot," or a "scatterdiagram," or a "scattergram." (In a moment we'll look at some examples.) Its function is to enable you to see visually how much scores on one dimension tend to parallel scores on the other. Scattergrams form a fundamental basis for the development of some of the more important properties of correlational data.

Let's suppose that we have a physical education class in junior high school comprised of 30 boys. For each one of these 30 boys we have taken a measurement of his height, and another of his weight. If we want to study these independently, we could form a distribution for, say, height, and study its plot, compute its mean and standard deviation, and make the appropriate statements about this group of boys as regards height.

Similarly, we could perform comparable operations with respect to weight. However, in order to find out something about the *relationship* between height and weight, we would start with a scattergram displaying these data *jointly*. Table 6-1 shows the measurements taken for each of the boys. (Please note that I have made up these data to illustrate our discussion. Thus, although illustrative, they are not necessarily typical of junior high school boys.)

Now we can and do plot and work with joint distributions of the raw scores like the one shown in Table 6-1. However, I would like to shift to grouped frequency distributions in order to facilitate the points I want to make. In Table 6-2, I have shown the grouped frequency distribution for each of these two variables separately.

Table 6-1. Raw score distribution of height and weight

Boy No.	Height (Inches)	Weight (lbs.)	
1	74	210	
2	72	180	
3	72	190	
4	71	165	
5	70	170	
6	70	140	
7	70	150	
8	69	200	
9	69	190	
10	69	140	
11	69	175	
12	68	160	
13	68	155	
14	68	140	
15	68	170	
16	68	150	
17	67	150	
18	67	145	
19	67	145	
20	67	160	
21	66	125	
22	66	130	
23	66	125	
24	66	130	
25	65	125	
26	65	120	
27	64	135	
28	62	120	
29	61	120	
30	60	111	
Y-Mean	67.5	(Height)	
Y-SD		3.10	(Height)
X-Mean	150.9	(Weight)	
X SD		25.51	(Weight)

Finally, Table 6-3 shows the same 30 students distributed jointly again, but this time *by using the midpoints* of the grouped frequency distributions to represent the scores in each category.

Again, looking at the joint distribution in Table 6-3, you can see which of the height measurements goes with each of the weight measurements and vice versa. You can also see that the use of grouped frequency data, in

Table 6-2. Grouped frequency distributions for height and weight

Height (Inches)	Mdpt.	f
73–74	73.5	1
71–72	71.5	3
69–70	69.5	7
67–68	67.5	9
65–66	65.5	6
63–64	63.5	1
61–62	61.5	2
59–60	59.5	1
		30

Weight (lbs.)	Mdpt.	f
201–210	205.5	1
191–200	195.5	1
181–190	185.5	2
171–180	175.5	2
161–170	165.5	3
151–160	155.5	3
141–150	145.5	5
131–140	135.5	4
121–130	125.5	5
111–120	115.5	4
		30

the form of the midpoints for each of the scores in each interval, makes for a lot of repetition of score values. Also, it is true that the grouping process reduces some of the precision in the measurements. Looking back at the distribution of ungrouped scores, and comparing the means and SDs with these, you will note minor differences that are the result of errors introduced by the use of category midpoints rather than the actual scores. However, these differences are quite small.

In any case, we will proceed to use this joint, grouped distribution as a basis for our discussion for the next few paragraphs.

First, as suggested earlier, we will plot the two distributions *jointly*. If we do this, using here the grouped distributions shown together in Table 6.3, we get a "scatterplot" that looks something like that shown in Figure 6-1. (Note that although all 30 boys are included in this and subsequent graphs, there are only 19 dots because some of them overlapped each other in the graphing.)

Several things might be noted about this figure. First, it makes plain that there is a distinct tendency for the heavier boys to be taller as well. Also, it

Table 6-3. Joint distribution of height and weight, grouped scores

Boy No.	Height	Weight
1	73.5	205.5
2	71.5	175.5
3	71.5	185.5
4	71.5	165.5
5	69.5	165.5
6	69.5	135.5
7	69.5	145.5
8	69.5	195.5
9	69.5	185.5
10	69.5	135.5
11	69.5	175.5
12	67.5	155.5
13	67.5	155.5
14	67.5	135.5
15	67.5	165.5
16	67.5	145.5
17	67.5	145.5
18	67.5	145.5
19	67.5	145.5
20	67.5	155.5
21	65.5	125.5
22	65.5	125.5
23	65.5	125.5
24	65.5	125.5
25	65.5	125.5
26	65.5	115.5
27	63.5	135.5
28	61.5	115.5
29	61.5	115.5
30	59.5	115.5

Y-Mean   67.4
Y-SD   3.10
X-Mean 148.2
X-SD   24.76

is clear that this tendency is far from uniform. As is true with most graphs, you can get an overall impression of the relationships (here between the height and the weight of these boys) much better from the plot than you can from reading the columns of data in the table. This is why scatter-grams are usually worth the time to construct—particularly if you don't know very much about the relationships being studied from prior sources.

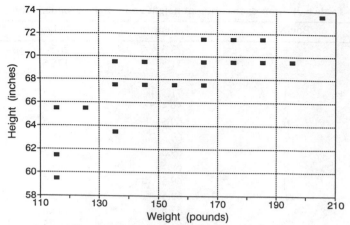

Figure 6.1. Scatterplot: Grouped Data

Well, all of that is interesting, you say, but so far it seems to be only an interesting way of assuaging one's curiosity about relationships. However, there are wider applications. Here is where the matter of prediction referred to above comes in. Suppose that you expected some new boys in your physical education class, and you had to make some decisions about uniforms, but you had no data whatsoever on the height and weight of the newcomers. Could you make a decision that would be better than chance? Of course you could!

IV. Prediction (Estimation)

One of the classic values of statistics is that it contains procedures for estimating—that is, predicting—unknown values, given that some relevant information is known. So, we shall now look at how statistical procedures can assist us in solving the problem of what size uniforms to buy for our new classmembers.

A. Estimation from a Mean

Any decision that is informed by facts (assuming they're true facts) is better—at least to some degree—than a chance decision. I could suggest all sorts of sources of information, but at this point I want to focus on one particular one as we develop the notion of prediction. That one is—as usual—the average (or mean). *If you had no other information, but did know the mean height and mean weight of last year's class, your statistically best estimates for your newcomers would simply be to assign those means to each incoming boy*! (A mathematical proof of this assertion is possible, but would take us beyond the scope of our concerns here.)

You will recall that in most cases (that is, where there exists a central tendency), the mean is the best single value to represent a distribution (it is most like more of the individuals than any other measure). So, in the total absence of other information, the value to use for any unknown case is the mean of the group to which the unknown case belongs.

However, it is important to note that the value of the mean as a predicted value for an unknown case improves if we can narrow the group upon which it is based to a more and more similar set of individuals (increase the central tendency by excluding uncharacteristic cases). In the present instance, for example, if you wanted to use a mean height to represent the height of a newcomer, you would not use a mean height for all people. It would be better to use a mean for boys; better yet to use a mean for boys of the same age as the newcomer; better yet, perhaps, to use the mean for boys of the same age from last year's class—and so on.

Now, to extend this line of thinking, suppose that you have collected information from each member of your class as to their heights and weights, and that (for whatever reason) you are able to measure the weights of the newcomers, but not their heights. Do you suppose that you could use your information on weight to improve your estimates of their heights?

B. Prediction Based on a Second Variable: Regression

The answer, of course, is yes. How accurate those estimates might be is another issue, but we have observed that the scatterplot suggests that the two variables are related. Whenever this is the case, a better-than-chance prediction is possible.

Suppose one of your new boys weighs 133 pounds. What would be your best guess for his height? Well, you could simply assign him the mean height of your class. But, you can improve your estimate by drawing on the data you have collected for your class on height and weight, and the fact that height and weight are related.

You have four boys in your class whose weight is in the same category as the newcomer (look back at the joint, grouped frequency distribution to verify that this is true). A better estimate might be obtained by assigning the newcomer *not* the mean height of the whole class, *but rather* the (more restrictive) *mean height for those four boys in his own weight group*.

Looking at your joint distribution, you see that the four boys in the 131–140 pound range have heights of 63.5, 67.5, 69.5, and 69.5 inches (the midpoints of the intervals in which each one is found). Averaging these four gives you 67.5, which you assign as the estimated height for your new boy.

The argument for this is of course that generally the mean is the most "typical" single value to represent a group of numbers. But, *if two variables (here height and weight) are related*, the boys with similar weights will tend to have similar heights. Therefore, the mean of those boys in the same weight category as the newcomer will be a superior estimate to that of the mean of all of the boys. How much superior obviously depends on how strongly the two variables are related.

Now, if one were to extend this argument to cover many incoming boys, one would soon have calculated the mean height for each of the 10 categories of weight in the distribution. Although one might use this set of 10 height means, as is, to make estimates for height given a known weight, both height and weight are commonly thought of as continuous variables.

A continuous relationship between the two can be approximated by passing a "straight line of best fit" through the 10 points representing the height means. This is permissible whenever the points appear to distribute themselves in a reasonably linear fashion. (It is generally accomplished by using a mathematical technique for line fitting involving the minimizing of the squares of the differences between the points and the fitted line, taken over all points.)

An additional advantage of adopting such a procedure as this is that it enables us to work directly with raw data. Rather than having to group the data first in order to calculate column means, the pairs of raw data can be plotted directly on the scattergram and the appropriate line drawn through these points. (I will present an example based on raw data later.)

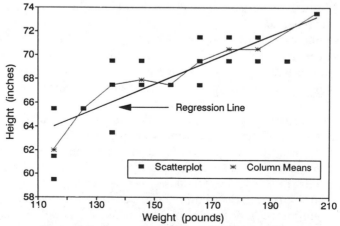

Figure 6.2. Regression of Y on X: Grouped Data

Because this is the way it is usually done, and because it provides a basis for the development of the index of relationship in a way that we will look at later, we will use a continuous line of best fit in the rest of our discussions in this chapter. Look at the column means that I have marked on Figure 6-2. If, instead of just connecting them, which would give a graph with a zigzag pattern, I draw the single straight line that best fits them, I get the graph shown in the figure. (I have put both lines on the graph to illustrate my point.)

1. The regression line

To repeat, the "straight line of best fit" functionally is simply a straight line arranged so that it passes through a set of points in a way that mathematically *minimizes* the amount by which it misses the points, taken as a whole. Obviously, the closer that the points in a scatterplot approximate a straight line, the better the fit will be (and the less error you will have in estimating one variable from the other). (As we will see later, the "goodness-of-fit" of the line provides another way to develop the index of relationship we have already talked about.)

The straight line that we have just identified (a line of best fit through the height means of the weight categories) is called a "regression line." More particularly it is the "linear regression of height on weight." In simple language it is a line representing our best prediction of height, given that we know weight. To use it, one takes any weight, finds it on the horizontal scale, then proceeds directly upward (vertically) to the line. That point on the line corresponds to the estimated (predicted) height,

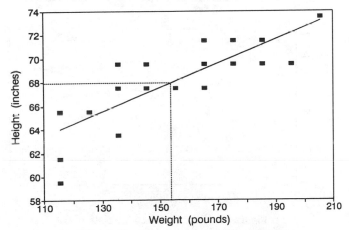

Figure 6.3. Prediction of Height from Weight

which is found by reading directly across (horizontally) from the point to the height scale.

I have illustrated this procedure with dotted lines in Figure 6-3 for a boy of 154 pounds. There you can see that his predicted height would be a little under 68 inches.

Of course, it isn't necessary to do all of this plotting, although, as with graphing in general, it is usually instructive to examine the picture of the relationships one is studying. Both the regression line and the value of any estimated score can be calculated mathematically without the use of graphic techniques. The formulas are very simple and are presented later.

2. Examples of regression lines

To give you insight into the way various data result in scattergrams representing various levels of relationship, I have included graphs for several sets of artificial data (made up for the purpose).

Figure 6-4 represents a relatively low positive relationship. There is some shape to the swarm of points, but obviously the line of best fit falls quite far away from some of them. Prediction from these data is not very dependable.

In Figure 6-5, we have an example of "independence," or no relationship. In effect, prediction is no better than chance, and the swarm of points is essentially circular, showing random ys for any x.

In Figure 6-6, the opposite is true. Here, the relationship is so strong that the points are almost all virtually on the line. Prediction of y from x is practically perfect.

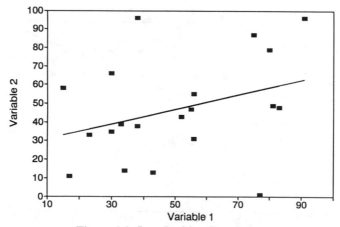

Figure 6.4. Low Positive Correlation

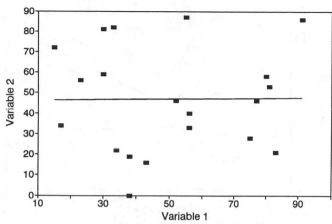

Figure 6.5. Zero Correlation (Independence)

Figure 6-7 shows a rather strong positive relationship, though not as strong as the one in Figure 6-6.

In Figure 6-8, we have a moderately strong relationship, but note that the direction of the regression line is different. Here high scores on the *x* variable go with *low* scores on the *y* variable. Thus we have an *inverse* relationship. Figure 6-9 shows another inverse relationship, of substantially less strength.

It is important to remember that direction of relationship and strength of relationship are two separate issues. The former is indicated by the direction

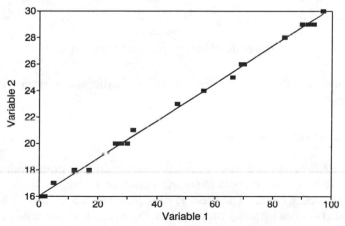

Figure 6.6. Very High Positive Correlation

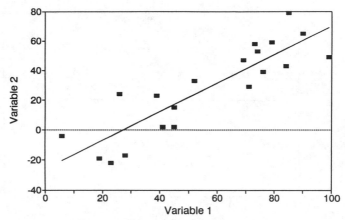

Figure 6.7. Strong Positive Correlation

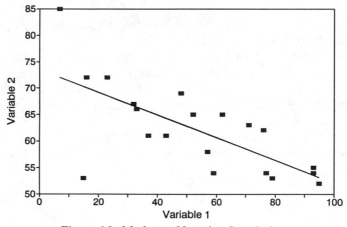

Figure 6.8. Moderate Negative Correlation

of the line, and the latter by how closely it fits the swarm of points on the scatterplot.

3. The other regression line

So far, we have talked only about predicting height given weight, or *y* from *x*. It is of course possible to talk about predicting *x* from *y*, or the "regression of *x* on *y*." The logic of your problem will indicate whether or not prediction is a one-way process. For example, if we were talking about the relationship between SAT scores and college success, prediction

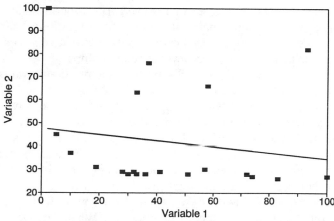

Figure 6.9. Low Negative Correlation

makes sense in only one direction. We would be interested in what test scores might tell us about college success—we would hardly use measures of college success to predict test scores! On the other hand, considering our height and weight example, it is just as reasonable (depending on circumstances, of course) to predict weight knowing height as it is the other way around.

You might be surprised, however, to learn that (with two exceptions) the regression line, and thus the predicted relationship, is different. It is *not* just a matter of reading from the *y*-axis to the same regression line and down to the *x*-axis, that is, reversing the process. In other words, you don't get the same results if you predict height from weight as you do if you predict weight from height, even though both predictions are based on the same underlying relationship derived from the same data set. An entirely different regression line is needed.

You will recall that the usual prediction, of *y* given *x*, employs a regression line that we originally established by grouping the data, computing the *y* means for each category of *x*s, and running the line of best fit through these column means.

The *x* on *y* regression line is established in an analogous way. We compute the *x means* for each *row* (i.e., *y*) category, and the regression line is the line of best fit through these *row means*. (In the case of ungrouped data, of course, the mathematics does essentially the same thing.) Then, for any known *y*, the predicted *x* is obtained by reading across the row to the line and then down to the corresponding *x* value.

Under what circumstances do you suppose that the two lines are the same? If you think about it a bit, you can see that where the relationship is a perfect one, and there is no error involved in estimating one from the other, each possible x and y pair must have a fixed relation to each other, no matter which way you go. This occurs only in two cases—for perfect direct relationships or for perfect inverse relationships.

[When there is absolutely no relationship the predictive value of the regression is of course zero no matter which way you go. This then becomes the general case where the best prediction you have of y is simply the general y mean (i.e., the mean of all of the y scores); and the best prediction of x is the general x mean. In terms of regression lines, the line of y on x is a horizontal line through the general y mean, and the line of x on y is a vertical line through the general x mean.]

In any other case you are in effect predicting ys from the y means of the x categories which involves errors based on how well those y means represent the ys in their respective categories. When you go the other way, you are predicting xs from the x means in each y category.

It is true that the total amount of error in the system is fixed by the strength of the underlying relationship found in the data set, and by the fact that the same xy cross-products are used in both calculations. However, the errors in predicting y from x are distributed differently than those in predicting x from y, because any given x mean may be either a better or a worse representative of its set of x scores than its corresponding y mean is of its set of y scores.

Figure 6-10 presents the y-on-x relationship which we have seen before for the height and weight data we have been studying. However, in addition I have plotted the x-on-y regression line to illustrate the difference between the two. They may or may not lie pretty close to each other,

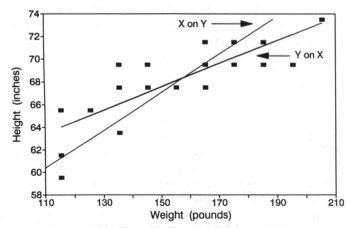

Figure 6.10. Y on X, X on Y: Height vs. Weight

depending on the strength of the relationship, as well as on the scales used to plot the two variables.

To reiterate, the goodness of either prediction is a function of the strength of the relationship underlying the predictions, which is the same whichever way you do the prediction; it depends only on the total amount of error in the system. But, the regression lines will be different for predicting x from y as compared to predicting y from x. Figure 6-10 illustrates, based on the same data as we used to show the line of fit for the ys, what is a rather typical set of regression lines for variables that are rather closely related. Note that the two lines differ rather considerably for these data.

In summary, the prediction technique outlined above is quite general. With it you can predict an unknown value on a particular variable from a known value on a related variable, given that you have sufficient data from similar cases to establish the regression line. The goodness of the prediction is a function of the closeness of the relationship between the predicted variable and the predictor variable. Visually, this closeness is indicated by the degree to which the plotted points are close to the line of best fit among them.

C. The Correlation Coefficient from a Scatterplot

Let's go back to our original data on the 30 boys and their heights and weights to look at several related topics, including how we get the index of relationship out of all of this "regressive behavior." I have presented both the raw data and that based upon grouping earlier in this chapter. Now, however, I want to consider these data in "standard distribution" form. You will remember that this refers to the distributions converted into a form where the means are zero and the standard deviations are 1 (also known as z score form). Scores for our 30 boys are presented again, including their z scores, in Table 6-4. In addition, there is some other information that will be explained later.

To restate, the joint distribution shown in Table 6-4 first presents again the data with which we are dealing, but taking them from the grouped distributions presented earlier (that is, using the interval midpoints as the scores). In addition to the scores in midpoint form, columns 3 and 4 in the table show the height and weight scores transformed into standard distribution ("standard deviate") form.

The scores in a standard distribution are often called "standard deviates," designated with a "z." I remind you again that this results from the application of constants and produces distributions with means of 0 and SDs of 1. The means and SDs have been calculated, and are shown at the

Covariance and Correlation

Table 6-4. Standard score and regression data

Boy No.	Height	Weight	z Ht.	z Wt.	$z \times z$	Estimated Ht.
1	73.5	205.5	1.97	2.31	4.554	1.8801
2	71.5	175.5	1.32	1.10	1.458	0.8957
3	71.5	185.5	1.32	1.51	1.992	1.2239
4	71.5	165.5	1.32	0.70	0.924	0.5676
5	69.5	165.5	0.68	0.70	0.473	0.5676
6	69.5	135.5	0.68	−0.51	−0.347	−0.4167
7	69.5	145.5	0.68	−0.11	−0.074	−0.0886
8	69.5	195.5	0.68	1.91	1.294	1.5520
9	69.5	185.5	0.68	1.51	1.021	1.2239
10	69.5	135.5	0.68	−0.51	−0.347	−0.4167
11	69.5	175.5	0.68	1.10	0.747	0.8957
12	67.5	155.5	0.03	0.29	0.010	0.2395
13	67.5	155.5	0.03	0.29	0.010	0.2395
14	67.5	135.5	0.03	−0.51	−0.017	−0.4167
15	67.5	165.5	0.03	0.70	0.023	0.5675
16	67.5	145.5	0.03	−0.11	−0.004	−0.0886
17	67.5	145.5	0.03	−0.11	−0.004	−0.0886
18	67.5	145.5	0.03	−0.11	−0.004	−0.0886
19	67.5	145.5	0.03	−0.11	−0.004	−0.0886
20	67.5	155.5	0.03	0.29	0.010	0.2395
21	65.5	125.5	−0.61	−0.92	0.562	−0.7448
22	65.5	125.5	−0.61	−0.92	0.562	−0.7448
23	65.5	125.5	−0.61	−0.92	0.562	−0.7448
24	65.5	125.5	−0.61	−0.92	0.562	−0.7448
25	65.5	125.5	−0.61	−0.92	0.562	−0.7448
26	65.5	115.5	−0.61	−1.32	0.809	−1.0729
27	63.5	135.5	−1.26	−0.51	0.645	−0.4167
28	61.5	115.5	−1.90	−1.32	2.514	−1.0729
29	61.5	115.5	−1.90	−1.32	2.514	−1.0729
30	59.5	115.5	−2.55	−1.32	3.366	−1.0729
Mean:	67.4	148.2	−0.01	0.00	0.812	
SD:	3.1	24.8	1.00	1.00		

Y on X (z form)		X on Y (z form)	
Regression Output:		Regression Output:	
Constant	−0.00965	Constant	0.0074
Std Err of Y Est	0.60085	Std Err of X Est	0.6016
R^2	0.6621	R^2	0.6621
No. of Observations	30	No. of Observations	30
Degr. of Freedom	28	Degrees of Freedom	28
X Coefficient(s)	0.8126	Y Coefficient(s)	0.815
Std Err of Coef.	0.1097	Std Err of Coef.	0.110

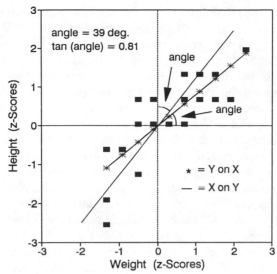

Figure 6.11. Height vs. Weight in Standard Scores

bottom of the table. Note that, as expected, the standard deviate (z) columns have computed means and SDs differing from 0 and 1 by only minor rounding errors.

The fifth column is the product of columns 3 and 4. Thus it is a column of "cross-products" of height and weight in standard deviate form. I have included it here simply to prove that the arithmetic mean cross-product (based on standard deviate scores) is indeed the correlation coefficient.

When we have developed this coefficient from our regression data, you will be able to compare the mean of this column, shown at the bottom, to the index of relationship derived from the regression data and note that they are the same. As noted earlier in this chapter, this index is usually given the symbol "r." I will come back to the last column in this table later, but it is a column of predicted (estimated) height values in standard (z) deviate form.

We now proceed with the development of the correlation coefficient. Figure 6-11 shows a scatterplot with both regression lines on it. It is based on the standard deviate scores for height and weight shown in columns 3 and 4 of the table. It also shows the lines of means (the vertical line is the mean weight, which of course is 0 in standard deviate distributions, and the horizontal line is the mean height, also zero for the same reason). *The lines of means and both regression lines always cross at zero in standard deviate distributions.* The zero point in such a plot is termed the "origin."

Figure 6-11, as well as the next few figures (6-12 through 6-15 and 6-17 and 6-18) also contain notations about "angles" and "rs" that I shall explain presently. They are intended to show how the regression lines move as a function of the strength of the relationship. These figures are done in terms of standard deviate scores (z-scores), but they are not based on real data, hence the plotted points are not shown. They show schematically how the lines are related. I have drawn a dotted reference line through the origin at a 45-degree angle, thus bisecting the northeast and southwest quadrants; I have drawn another dotted reference line at 45 degrees, bisecting the northwest and southeast quadrants.

For a relationship to be a positive one, high scores on one must tend to go with high scores on the other, and low scores must tend to go together as well. That requires that, for positive relationships, the regression lines run in the northeast and southwest quadrants, because according to the score scales, as shown, this can be true only for such a line.

Now consider what we already know about the two lines. First, if there is no relationship, the best prediction is simply the mean of the other variable; in other words, the regression line must coincide with the line of means. So, when $r = 0$ the y on x regression line must lie exactly on top of the y-mean (which in the case of standard deviate distributions as here is the horizontal axis). And, similarly, the x on y line must coincide with the vertical axis (the x mean). This condition is illustrated in Figure 6-12.

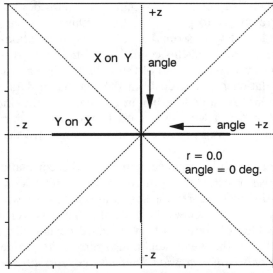

Figure 6.12. Regression Lines, Zero Correlation

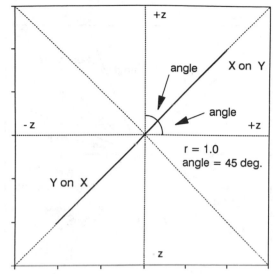

Figure 6.13. Regression Lines, Perfect Positive Correlation

We also know that when the positive relationship is a perfect one you must get the same result whether you go from x to predict y or the other way around. This has two implications: first, the lines must coincide with each other; and, second, they must coincide exactly at the 45-degree line, as this is the only position in which reading from the x-axis to the line to the y-axis gives exactly the same result as going the opposite way. Try this yourself. (See Figure 6-13.)

The next set of figures (Figures 6-14 and 6-15) illustrates the intermediate positive relationships. It can be seen that the positions of the two lines are mirrored across the 45-degree reference line, moving oppositely, proportionately, as the relationship drops from perfect toward zero. This is strictly true only when both distributions are in the form of z scores, although the direction of the movement is always this way.

Because in the standard deviate form, the regression lines go through the origin as a sort of pivot, the strength of the relationship is in proportion to the ANGLE between the regression line and the predictor variable axis. This angle is called the slope angle of the line. Note, now, that for the y on x line it is the slope angle with respect to the x-axis; *but*, for the x on y line estimates are made with respect to the y axis, and thus the relationship is proportional to the slope angle with respect to that line and the y-axis.

What is a slope? We can talk about it as an angle. For example, we sometimes talk about a grade or slope in a highway as being "about 5 degrees," or some such. But, we can also talk about it numerically, as

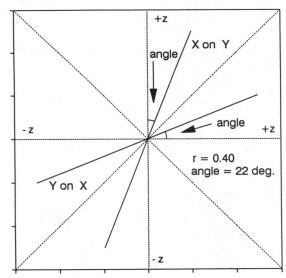

Figure 6.14. Regression Lines, $r = 0.40$

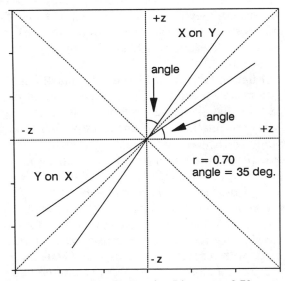

Figure 6.15. Regression Lines, $r = 0.70$

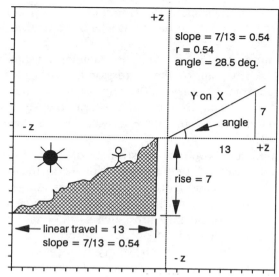

Figure 6.16. Slope Illustration

follows. Think about hiking up a hill. The slope is the steepness of the hill. Numerically, it is the number of feet of elevation you rise, divided by the number of feet straight forward you would have to move to get that much rise if the hill were not in the way.

If you think about the hill as a triangle, the path is the regression line and the elevation from the base to the top is the numerator of the slope. The denominator of the slope is the distance from the foot of the hill along a horizontal line under the hill to a point directly beneath the top. Figure 6-16 illustrates this point.

If we think of the figure as being measured in yards, we can see that the hiker must travel 13 yards horizontally in order to achieve the 7 yard rise necessary to gain the top of the hill. Thus, the slope of this hill is 7/13 or 0.54. (Reference to tables to be found in any trigonometry book tells us that this corresponds to a slope angle of about 28 1/2 degrees. You had better be a real hill-climbing hiker, if you want to travel this trail very far!)

The right hand half of the figure shows the analogous relationship for a regression problem. Note that slope is defined exactly the same way.

With respect to any regression line on any scatterplot, the slope angle is the angle that the line makes with the predictor axis. To determine this angle, you pick any point on the regression line to start with, and then drop a vertical line through it perpendicular (at right angles) to the predictor axis. Then the slope is the length of this vertical line divided by the distance from the origin to the point where the vertical line meets the axis.

Looking at our graph (Figure 6-16), for example, you should confirm this for yourself. Pick any point on the regression line; using a ruler, take the vertical measure from the x-axis to the regression line and divide it by the distance from the origin to the point where the vertical line meets the x-axis. You should get the same answer for the slope as is shown, approximately 0.54.

Now, I pointed out above that the strength of the relationship between the two variables scatterplotted is proportionate to the slope angle between the predictor axis and its corresponding regression line. Of course, angles are circular measures, marked off in degrees of a circle. It is hard to combine degrees with linear measures such as the ones we have used in our score scales, and marked off on our scatterplot, in any meaningful mathematical way. Therefore, when we try to express the strength of a relationship, we don't usually use the slope angle itself as our index of relationship.

However, because the *numerical* value of the slope as usually calculated (see above) is directly related to the size of the slope angle, we can use the numerical slope measure instead of the angle measure as an index of relationship. Again, remembering that we are talking about standard deviate distributions, *this slope measure IS the correlation coefficient, "r."* Go back to Figure 6-11 and use a ruler to determine the slope value for the y on x regression line. Now compare it to the mean cross-product shown at the foot of the z cross-products column in Table 6-4. They will be found to be the same, within errors of measurement.

Noting that the x on y regression line is exactly symmetrical in standard deviate distributions to the y on x regression line, you will find that r may also be calculated from the slope of the x on y line *with respect to the other (y) axis*, in the same way. You should apply your ruler to the graph in Figure 6-11 again to convince yourself that this is true. Also, check for yourself the last few figures presented.

(I again caution, however, that slope translates directly to the correlation coefficient *only* when the distributions are in the z score form. Otherwise, it is related, but only after the disparate means and standard deviations of the two distributions are balanced out.)

So much for positive relationships. How about negative (inverse) ones? Exactly the same situation prevails, except that now we must deal with the two lines in the opposite (northwest and southeast) quadrants. Figures 6-17 and 6-18 provide two illustrations of negative relationships.

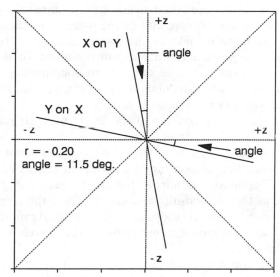

Figure 6.17. Regression Lines, $r = -0.20$

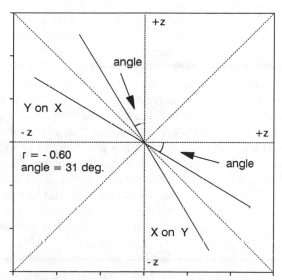

Figure 6.18. Regression Lines, $r = -0.60$

Note that in the inverse situation, the slope measurement results in a minus quantity, as in the northwest quadrant the numerator of the slope fraction would be at the positive end of the y-axis, but the denominator would be on the negative side of the x-axis. In the southeast, the y measure would be negative and the x measure positive. In either case the slope fraction would result in a negative, corresponding with the fact that for regression lines in these quadrants the prediction is that high scores on one variable go with low scores on the other.

In all other respects the position of the lines and their relationship to the slope angles and their numerical values is entirely analogous. For instance, the two lines coincide at the 45-degree line running northwest and southeast when the relationship is a perfect negative one (-1.0).

Again, the lines move (in mirror fashion to each other across this 45-degree line) as the relationship weakens, reaching the respective mean lines (x- and y-axes) as the relationship falls to zero. Again the slope of a line against its axis, measured numerically, is the correlation coefficient (a negative value of course).

It is important to remember that the correlation coefficient is numerically equal to the slope measure directly *only* when the problem is set up in terms of the z scores for both variables.

[Those of you who are familiar with trigonometry will recognize that the measure that I have termed the slope fraction is in reality the tangent of the slope angle. The range of the angle is from plus 45 degrees to minus 45 degrees, and of course the tabular value of the tangent for this range is from $+1.0$ through 0.0 (at 0 degrees) to -1.0—corresponding with the expected range for the correlation coefficient.]

I have been careful to qualify the above discussion by reminding you that the slope is exactly the correlation coefficient only for standard deviate distributions. In fact, the basic relationship holds for all joint distributions, *but* if the standard deviations are different for the two variables, the slopes of the regression lines are affected proportionately.

In such cases, in order to get the correlation coefficient, the effect of the disparity in SDs must be adjusted by multiplying the slope obtained by the ratio of the two SDs in such a way so as to cancel out the effects of their difference. (This is a straightforward process, but can be confusing to the layman, so I will leave the matter at that, because we rarely actually calculate correlation coefficients in this manner.)

One final point is that there is obviously a mathematical relationship between the slope of the regression line and the mean cross-product, as each of these yields the identical correlation coefficient. However, we need not go into the mathematics here.

V. Applying the Prediction Process

I have noted earlier that the basic process of prediction when done graphically consists of: collecting data on two variables for a group of individuals or items; plotting the data set pairs on a two dimensional graph; drawing the straight line of best fit through these points (either with respect to predicting y from x or x from y or both); taking a known value on the predictor scale, reading perpendicularly to a point on the appropriate regression line, and then referring from that point perpendicularly to the second scale, where the estimated value of the second variable may be read.

I have also mentioned that this process may be accomplished without using graphic methods. Although it is not my intent to go into all of the variations on this, I think it is worth our while to consider the simplest form of prediction using that dreaded word "formula".

The secret of the prediction is obviously the regression line. This is because the regression line represents the predicted values of the missing variable. Think about the graphic process. If the known values of x are read on the baseline, we then go up to the regression line to find the predicted value of y. Thus, the line represents all possible pairs of a known x and its corresponding predicted y. Or, conversely, all predicted ys lie on the regression line!

So, we start with the fact that any straight line has a simple formula. If we confine ourselves for the purposes of this discussion to the task of predicting y from x, then the regression line is:

$$\tilde{y} = bx + a.$$

In this equation, x is the known value of the predictor, and $\tilde{y}$ is the estimated value of the unknown y. The constant b is the slope of the regression line (which, as we have just discussed, is related to the strength of the relationship between the two variables). The a is another constant that relates to where the regression line crosses the axis of the variable to be predicted (here, the y-axis).

If we deal with raw scores, this standard equation for a line is expanded to take account of the fact that the two distributions have different means and standard deviations. If we eliminate the effect of different means by converting the distributions to standard scores each with a mean of 0, the a goes away, but we still have to correct for the different standard deviations by multiplying by the ratio of the standard deviations, as explained earlier.

If we convert the two distributions into standard deviate form, no corrections and adjustments are necessary, b is simply the correlation coefficient and the equation reduces to:

$$\bar{y} = r(x).$$

Although this $\bar{y}$ is correct, it is of course in the same dimensions (units) as the standard deviate distribution. If you wish to contemplate it in the original units, you must reconvert each $\bar{y}$ by multiplying it by the original y standard deviation, then adding the original y mean.

To illustrate, look back at the last column in Table 6-4. There you see the predicted heights, in standard normal form for our 30 physical education boys. To get back to heights such as those in column 1, you would have to convert these predicted heights back to scores with the original mean and SD. This is easily accomplished according to the principles we went over earlier in this chapter for applying constants to change the statistics of a distribution.

Remember that the standard deviate form is one where the mean is 0, and the SD is 1. To convert any scores back to the statistics of the original height distribution, you would first multiply the standard deviate value in the last column by the original SD for height (3.1), then add the height mean (67.4).

For example, the estimated (predicted) height for boy number 10 would be $(-.4167)(3.1) + 67.4 = 66.1$ inches tall. For comparison, the actual height of the 10th boy was 69 inches (refer to original, raw score distribution), and his height in the grouped distribution was 69.5 inches (the midpoint of the interval 69–70, which is really 68.5–70.5 according to our procedures for grouping scores, as discussed in Chapter 3).

Thus, the estimate would appear to be off quite a bit—but this is only to be expected, since this boy was rather thin compared to the others in this group. This illustrates the fact that although the relationship is strong overall, individual predictions may be rather poor.

VI. How Good is Our Estimate?

So far, we have focussed on the method of getting an estimated value. No consideration has been given to how far the estimate may miss its mark. In this section we will consider this very important question. Doing so not only gives us a way to think about and to state how good our estimates are, but it also provides us with one more way to define the coefficient of relationship, "r."

Let's think about an individual's height score in relation to his or her weight score and the y on x regression (in standard deviate form) that we have been using as our discussion data. If y is his or her measured score, we can think about this y as being comprised theoretically of three pieces: the y mean (here zero), plus a predicted score $\tilde{y}$, plus some amount (either positive or negative) by which the prediction missed the mark (technically called a "residual").

To see this more clearly, look at our raw data scatterplot, which I have reproduced in Figure 6-19. Our first guess for the actual y corresponding to a given x would be the y mean; with regression we would have a $\tilde{y}$ piece, either above or below the y mean; then, as compared to the actual y value, this predicted y is off its mark by the amount of the residual (again either plus or minus).

Taken across the group of 30 boys, the y or height mean does not vary —it is the same for all boys; however, the predicted heights (which are the values that lie on the regression line of y on x) vary above and below the y mean, and the amounts (residuals) by which the line missed the actual height measurements can be either positive or negative (depending on whether the actual y scores were above or below the regression line).

Now, let's think about the variance in the height scores in our set of data. (Remember, variance is the same as the square of the SD.) As mentioned early in the last chapter, the variance is construed as a measure of the "total tendency of the scores in a group to differ among each other." The y-variance in our example (again, to simplify, using N instead

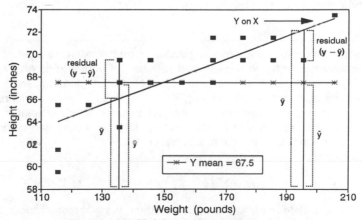

Figure 6.19. Height vs. Weight, A Residual

of the more correct $N - 1$) is:

$$V_y = (y - M_y)^2/N.$$

This value represents the total height (y) variance in this set of data. However, this total variance can be divided into two discrete pieces, one based on the variation of the predicted scores from the mean, and one based on the extent to which the predicted scores missed the real scores (the variance of the residuals from the regression line). [I will not attempt to argue the proof of this statement here.]

In other words, the total height variation in the system is made up of the "successfully predicted" (sometimes called "related" or "common") variance plus the "not-successfully predicted" or "error" variance. This division of the variance could be represented as follows:

$$V_{\text{total}} = V_{\text{error}} + V_{\text{common}}.$$

If we wrote this out in terms of scores, it would be:

$$\sum (y - M_y)^2/N = \sum (y - \bar{y})^2/N + \sum (\bar{y} - M_y)^2/N.$$

I have continued the development started above in the box, so that those of you who do not wish to wrestle with the algebraic operations involved may continue on. However, for those of you who are interested, the material in the box shows how we get to the following statement, which is exceedingly important in the matter of interpreting correlation coefficients:

$r^2 = $ *Proportion* of total variance that is predictable!

Conversely, the *proportion* of the total variance (of y) which is "error" (nonpredictable from this relationship) is: $1 - r^2$.

[Most statistics books would now drag you through the extensive algebra and manipulation necessary to expand and reduce the above to its final form. However, we will pull a fast one, and simply assume that we are again working with *standard deviate scores*. We will, in time, arrive at a final result that is applicable to all distributions; it's just that it is much simpler to get there it if one doesn't have to deal with the differing means and standard deviations and their effects on the arithmetic!]

We start with:

$$\sum (y - M_y)^2 / N = \sum (y - \bar{y})^2 / N + \sum (\bar{y} - M_y)^2 / N.$$

Looking at the three terms in this equation, then, the first term reduces to 1. (Recall that the SD of a standard distribution is 1, and thus its square, the variance, would also be 1.) The second term we will leave alone. It is called the "error variance" or the "error of estimate." It is of course the arithmetic mean of the squared residuals, as you can see above. The third term we must deal with. Earlier we learned that in a standard deviate distribution $\bar{y} = bx$; we also know that in such distributions the mean is zero. Substituting these values in the third term above gives us:

$$\sum \left(\bar{y} - M_y\right)^2 / N = \sum \left(bx - 0\right)^2 / N.$$

Squaring and summing, this becomes:

$$\sum \left(bx - 0\right)^2 / N = b^2 \left(\sum x^2 / N\right).$$

However, we remember that in standard normal distributions $\sum x^2 / N$ (the SD) is 1. Also, in such distributions, the constant, b, is the correlation coefficient, r.

So, then, the third term reduces to r^2! Now, let's see where we have come out. Our original expression showing that the total variance is the sum of the error variance and the common or predicted variance has reduced to the following:

$$1 = V_e + r^2.$$

This is an important statement showing the relationship of the predicted variance (as represented by the squared correlation coefficient) and the error variance. However, it is in need of some adjustment to make it applicable to distributions in general, because we developed it on the basis of the special case of standard deviate distributions.

Remember that the 1 in this equation was originally the variance of the y variable. Suppose we first restore the symbol for the y variance in place of the 1; and, then suppose we divide each term in the equation by that same variance. It would then look like this:

$$V_y / V_y = V_e / V_y + r^2 / V_y.$$

The first, and perhaps most important, aspect of this last statement is that it provides a way to think about goodness of prediction, or strength of relationship, in terms of percentages. The equation now shows the breakdown between the two components of variance in terms of their proportions to the total.

Thus, this statement reveals that the *proportion of the total variance accounted for by the relationship between x and y is r²!* This is an exceedingly important finding because it offers a convenient and easily understandable way to describe the extent of the relationship between two variables.

For example, in the case of our height predicted from weight example, the correlation coefficient, r, was approximately .812. Squaring it we get .659.

This tells us that 65.9% of the variance in height was predictable from weight! Other ways to state this same fact would be: 65.9% of the variance between the two variables was "common;" or, the strength of the relationship was on the order of 66 out of 100.

CAUTION: This statement has no meaning for any particular individual (statistics apply to groups); and, it most certainly does *not* mean that weight *causes* height, or vice versa.

As noted, the statement also allows us to express the proportion of error, which, of course is simply what's left. Thus, if r^2 is the proportion of predicted variance, the proportion of error variance is simply $1 - r^2$!

Now, it is important to recognize that this is only a statement of proportion. If it should be desirable, and it often is, to state the error variance in terms of the original units of the distribution, you simply take the variance of the distribution, V_y, and multiply it by the proportion which is error:

$$V_e = (V_y)(1 - r^2).$$

This statement sets forth the error variance in terms of the original height distribution. In our example, this would be (9.59) (.341), or 3.27 square inches. To make more sense of this, of course, it would be better to talk about the standard deviation, so that we don't have to worry about square units. Recalling that the SD is simply the square root of the variance, and taking the square root, we have the average error as 1.81 inches.

This last relationship is often expressed generally in the literature as the "standard error of estimate." Thus, in the more common form we would take the square root of both sides of the above equation, which yields the following form:

$$\text{SD}_e = \text{SD}_y\left[\sqrt{(1 - r^2)}\,\right].$$

This statement says that the standard deviation of the errors (residuals) in predicting y from x is equal to the standard deviation of the y distribution times the square root of one minus the squared correlation of variables x and y. This result is in terms of the original units—inches of height.

As noted, this term SD_e is often called the "standard error of estimate." The expression $(1 - r^2)$ is the proportion of error variance associated with the prediction. As mentioned earlier, r^2 is the proportion of common or predictable variance in the relationship between x and y. Remember that r, and of course r^2, are the same whether you are predicting y from x or vice versa. This is logical, because the goodness of prediction is a function of the strength of the relationship, no matter which variable is the predictor variable.

There is also a name for r^2—it is sometimes called the coefficient of determination.

A. The Correlation Coefficient Redefined

Note, now, that these relationships have given us one more way to define the correlation coefficient—in terms of the errors of prediction. If we restate the last variance equation in terms of r, it would be:

$$V_e = V_y(1 - r^2), \text{ so}$$

$$r^2 = (V_y - V_e)/V_y, \text{ and}$$

$$r = \sqrt{[(V_y - V_e)/V_y]} = \sqrt{(1 - V_e/V_y)}.$$

[In passing it should be noted again, that if one had the time, space, and inclination, it is possible to show that all three of our approaches to the definition of r (mean of the cross-products, tangent of the regression angle, and function of the errors of prediction) are mathematically equivalent.]

Now, finally, we are in a position to explain the mysterious information presented along with the scores in Table 6-4, under the heading of "Regression Output." These data were produced by the computer program Lotus 1–2–3, one of many available today that can be used to perform what manually would be tedious calculations necessary in the correlation/regression process.

The "X-Coefficient" is the slope fraction we have been discussing. In problems such as ours where the calculations have been based on scores in the standard deviate form, the X-coefficient is the same as the correlation coefficient, r. (However, I should emphasize that all of the computed data I have presented in this chapter may be subject to rounding errors which I have introduced at various points in simplifying the numerical data.)

The value labeled "Constant" is the "a" in the regression line equation presented earlier. In any distribution a predicted y is obtained by multiplying x by the X-coefficient and adding the constant ($\bar{y} - bx + a$). Notice also that in this case, because of the standard deviate form, the constants are zero (within rounding error). This is because when the means equal zero, both regression lines run through zero, at the origin where the mean lines cross—just as we noted in our graphical analysis.

r^2 is the square of the correlation coefficient, or the coefficient of determination, and "Std Err of Y Est" is the standard deviation of the residuals or the "standard error of estimate for y predicted from x."

We have not talked about the "Std Err of Coef," which pertains to the statistical stability of the X-coefficient, as it goes beyond our purpose here. Similarly, I do not intend to discuss the topic of degrees of freedom, as it is not needed to understand the fundamentals with which we have been dealing.

The second set of regression outputs presents the analogous data for predicting x from y. Note that, as it should be, r^2 is the same as it was for predicting y from x.

I have taken the trouble to present these computer output data only because the audience for a noncomputational book, such as this one, will most likely acquire any correlational data it uses through one or more of these computer sources. I simply wanted you to see what such output looks like.

VII. Summary

We have covered a lot of ground in this chapter. I first tried to establish the similarity of variance and covariance. In a sense both are concerned with the tendency of scores in a distribution to differ among themselves. The variance assesses this tendency within a single distribution, whereas covariance assesses it jointly for two distributions. In both cases, the most familiar and useful indices are based on a multiplicative relationship, and we illustrated how multiplication can be used to show magnitude and direction of relationship.

Using the multiplication process, applied to "standard," or "standard deviate" distributions (means of 0, and SDs of 1), I developed an index of the relationship between two variables. It is called the correlation coefficient ("r"), and for such distributions it is the arithmetic mean of the cross-products (a cross-product is the multiplication of the two members of a pair of scores), taken over all of the "cases" in the distribution.

The correlation coefficient can take any of a range of values from $+1.0$ (perfect direct relationship) through 0 (independence, or no relationship) to -1.0 (perfect inverse relationship).

We digressed to discuss the effects of operating on a distribution with constants (a value applied to each score). Means are raised or lowered by the exact amount of an additive constant, whereas SDs are not affected; means are multiplied or divided by the exact amount of such constants, as are SDs. These facts were shown to be a basis for the construction of derived distributions having whatever means and SDs might be desired. "Standard scores" used in testing work, and "standard" distributions are developed this way.

Any distribution can be converted to a standard distribution by the application of two simple steps: first, subtract the mean from each score, and then divide the result by the SD. This produces a derived distribution that is standard—has a mean of 0 and an SD of 1. Symbolically: $z = (x - M_x)/\text{SD}$.

I described how we can use the relationship between two variables to predict unknown values on one if we have the corresponding values on the other. In the case of predicting y from x (the regression of y on x), the equation is: $\bar{y} = bx + a$, where a is a constant known as the y-intercept, and b is a constant based upon the relationship between the x and y variables.

If a set of paired scores in grouped form (one score for each of two variables) is plotted together on a graph, with x on the horizontal axis and y on the vertical, the regression of y on x is represented by a line drawn through the y means for each of the x categories on the baseline. In the case of continuous data, a line of best fit can be developed mathematically that does the same thing. It is this line that is called the regression line for predicting y from x.

In analogous fashion, a regression line for predicting x from y can also be developed. This line is tantamount to a line through the row means; in regression form it is called the regression of x on y. It is the same as the y on x line only when the relationship is perfect.

I showed that for two variables in standard deviate form, the "b" constant, above, is the tangent of the angle between the regression line and the predictor axis (for either y on x or x on y). For such variables, b is exactly equal to the correlation coefficient, r.

Finally, I noted that our discussion had not yet approached the question of describing the goodness of estimating y from x. To solve this problem we developed the standard error of estimate—which is the standard deviation of the residuals (the differences between the predicted ys and the actual ys). In doing this we found that "r" could be defined, for the third time, this time as a function of the residuals.

In dealing with the standard error of estimate, and its relationship with r, we uncovered a useful way to interpret the strength of a relationship between two variables. The value "r^2", also called the coefficient of determination, is the proportion of the total variance in the predicted variable that is common with, or predictable from, the predictor variable. Thus, a correlation of 0.3, "explains" only 9% of the variance in the predicted variable—the other 91% of the variation is either independent of the predictor, or a function of sheer chance.

VIII. References

Achen, Christopher. *Interpreting and Using Regression*. Beverley Hills, CA: Sage, 1982.

Agresti, Alan. *Analysis of Ordinal Categorical Data*. New York: Wiley, 1984.

Anderson, T. W. *An Introduction to Multivariate Statistical Analysis, 2nd Ed*. New York: Wiley, 1984.

Draper, Norman and Smith, Harry. *Applied Regression Analysis, 2nd Ed*. New York: Wiley, 1981.

Hoel, Paul G. *Introduction to Statistical Theory*. New York: Wiley, 1984..

Manly, Bryan F. J. *Multivariate Statistical Methods: A Primer*. New York: Chapman & Hall, 1986.

Srivastava, M. S. and Carter, E. M. *An Introduction to Applied Multivariate Statistics*. New York: North Holland, 1983.

7

More About Relationships

I. Overview and Purpose

Although the correlation coefficient that we discussed in Chapter 6 is the most popular measure of relationship used by statisticians, it is not the only one used. Circumstances may suggest that one or another of several possible measures are more appropriate. In this chapter I want to introduce you to several of these other measures, and describe the circumstances in which their use is considered desirable.

In addition, it may have occurred to you that prediction (regression) and relationships (correlation) are somewhat restricted if we are limited to just two variables at a time. This is true. Also, in fact, it is quite common to have several predictors combined in an effort to predict one criterion (predicted) variable. Therefore, we will also take a brief look at "multiple regression," as this case is called, along with the measure of the strength of that relationship, called the "multiple correlation coefficient" ($"R"$). (It is beyond our purposes to delve into the case where multiple variables are employed both as predictors and criteria.)

Finally, I will conclude this chapter with some comments and cautions on correlations, and on the subject of "causation" and its relationship to correlation in general.

II. Other Measures of Bi-Variate Relationships

We have spent a great deal of time on the subject of the most general case of the linear relationship of two variables in terms of regression and correlation. This has been justified, because the product-moment, or Pearsonian, correlation coefficient (r) is the "star" of such measures. There are, however, situations where for one or another reason r is not the appropriate index. We shall take a brief look at some of these in this section.

Most of these measures may be interpreted as variants of the standard correlation coefficient. Because the purpose of this book is familiarity with concepts and ideas rather than computation, I will not go into detail with respect to these measures. Should you need to compute such measures, I would recommend consulting a statistician. But on the basis of the material presented herein, you should have sufficient understanding to be an intelligent consumer of his or her work.

A. Rank Order Correlation

As we have noted in a previous chapter, some situations do not lend themselves to cardinal measurement—placing the data in rank order is the best that we can do. Does this mean that we cannot measure the relationship between two such variables? Not at all. The eminent English statistician, Charles Spearman, worked out a variant on the correlation coefficient, which has been named for him, Spearman's R.

Spearman's R is equivalent to applying the Pearsonian correlation to data comprised of ranks. Its interpretation is identical. But, it must be understood that, being based upon ordinal data, it utilizes less information about the variables being studied than the product-moment correlation would if it could be used. (Refer to our discussion of cardinal versus ordinal numbers.)

When calculating Spearman's R, it is necessary only to arrange the data in the usual pairs for the items or individuals in the study set. First, the items are ranked on the first variable, then on the second. You must make sure that the ranks run in the same direction for both variables. That is, if the first variable is ranked from most to least, with rank 1 for most, then the second variable must be similarly ranked. Both ranks must exist for each case in the calculation, or that item must be dropped.

If the ranking process should encounter any identical values in the distribution of values being ranked, the rank orders are split among the tied values. For example, if we get to rank 15 and encounter 3 identical values, then ranks 15, 16, and 17 are tied. In this case, we add up the ranks at issue and divide by the number of them, assigning the result to all such tied values. Here we would add 15, 16, and 17, getting 48, and divide by 3, which equals 16. Thus items 15, 16, and 17 would *each receive the rank of 16*, and we would resume with the next rank being 18. (Notice that we have again resorted to the mean—here the mean rank of the tied values.)

B. Biserial Correlations

It sometimes occurs that one of the variables exists only in terms of a dichotomous variable (one with only two possible values). An example would be "Did you vote in the last general election?" The possible answers here are "Yes" or "No." Is it possible to correlate such a two-choice variable with continuous variables of interest, such as family income? Yes, using another approximation of the product-moment correlation called the "point-biserial correlation." The point-biserial procedure is used in the instance where the dichotomous variable is assumed to be a true dichotomy, which is probably the case in this example.

Another biserial procedure applies, simply called the "biserial r," if the dichotomy is assumed to conceal a truly continuous variable. An example might be if we classified businessmen into two categories: Successful and Unsuccessful. It is probable that success in business is really a continuum, and that our dichotomy is an oversimplification. In addition to this, you should understand that numerically somewhat different results will usually be obtained, depending upon where on the continuum the break-point between the two portions of the dichotomy is placed.

Both the biserial and the point-biserial procedures provide approximations of the Pearsonian coefficient, under slightly different assumptions. Both are interpretable in the same way, although the biserial can produce values greater than 1 under unusual circumstances. Both are somewhat less powerful than the Pearsonian correlation, because dichotomous measurement is somewhat imprecise as compared to measurement on a continuous dimension.

C. Categorical Correlation

Another situation that sometimes arises concerns the correlation of *two* dichotomous variables. If they are true dichotomies, such as the voter question above correlated with gender, the proper procedure is the use of a "phi coefficient." This is the two-dichotomy analog of the point-biserial correlation mentioned above.

If the two dichotomies are artificial, the proper procedure is the "tetrachoric *r*." This is the analog of the biserial *r*, and it too may be influenced by the location of the dichotomy break-points. It also may yield strange values under extreme or unusual circumstances.

In the case of either the phi coefficient or the tetrachoric *r*, we are again approximating the Pearsonian *r*, and the interpretations are similar. Note, however, that both of these measures have diminished power because of the imprecision of dichotomies as compared to full-score distributions.

D. Nonlinear Relationships

Possibly you have noticed that I have assiduously ducked the issue of what happens if your scatterplot clearly shows that a straight line is not a good representation of the relationship between the two variables in a study. (Figure 7-1 shows an illustration of such a scatterplot.) This is not at all uncommon.

One example that comes to mind is the relationship between income and amount of education. It is true that data on the American population generally support the notion that the more education one has the more one can earn. But, historically, this has been true only up to a point; sometimes average earnings decrease as education increases. One possible explanation is that persons with advanced graduate degrees and postdoctoral training frequently teach, and teaching has historically been one of our lower paid occupations.

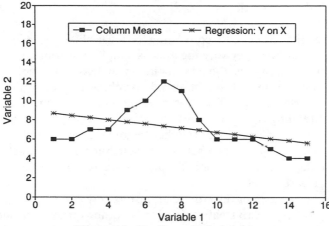

Figure 7-1. Nonlinear Relationship

I have not discussed the issue of nonlinear relationships because their treatment gets considerably more complicated mathematically than we can deal with here. However, let me describe a general procedure that can be applied. First, there must be a clear curvilinearity to your scatterplot. Then if you divided the predictor variable into a number of categories—as many as the data will support, you will be able to compute the column means of the y variable, just as we did in our regression problem. Here, however, you simply connect the column means, and the resultant, jagged line becomes your line of best prediction. (If you wish, the jagged aspects of the line can be partially smoothed by various forms of "curve-fitting," either graphically or mathematically.)

As you can see from Figure 7-1, passing a straight line through such means, using our standard regression line approach, provides a poor approximation of the column means, as compared to a line connecting them all. Similarly, your predictions from such a regression line would be poor ones as compared to using the line connecting the column means. *Where the true relationship is a curvilinear one, the correlation coefficient is always an underestimate of the true relationship*, because of the poor fit of the (straight) regression line to the (curved) set of points.

An approximate index of the strength of the curvilinear relationship in cases such as the one described above is called the "correlation ratio," or "eta." (For the computational formulas for eta and the other measures described in this section, it is best to consult one of the texts from the reference list for more detail.)

III. Correlative Analysis

One of the difficulties with correlating two variables in order to learn something about how they vary together is simply that some other characteristic that you have not thought about, or at least have not measured, may have more to do with the observed correlation between the first two then they do themselves. Take for example the problem of correlating scores on a reading test with scores on a math test. Arguably, both sets of scores are affected by a number of other factors that are not obvious in the simple correlation of the two sets of scores. Such factors might include general intelligence, socioeconomic status, years of schooling, age, education level of parents, etc.

To get a better idea of the true relationship between reading and math, we would probably try to treat some of these other variables in such a way so that their influence wouldn't be so pronounced on the correlation we are trying to assess. The effort to look at the correlation between two variables, with the effects of one or more other variables "held constant," or "partialled out" of the calculation, is called "partial correlation."

In effect, the correlation on reading versus math is computed on the covariance remaining after reading and math are each adjusted to eliminate the portion of its variance due to its covariation with an unwanted ("control") variable, such as intelligence. Thus, we would be trying to answer the question, "What is the relationship between math and reading scores, if any, over and beyond the effects of the fact that both of them are related to intelligence?"

If the problem is one of comparing mean performances of two or more groups on some study variable of interest (say, bushels per apple tree), perhaps under different conditions, a similar process is frequently used; it is referred to as "analysis of variance" (ANOVA). Again, we might wish to consider the effects of co-variation of the variable of interest with another, "control" variable. The procedure analogous to partial correlation is called "analysis of covariance (ANCOVA)."

In either case, the effect is to assess the covariance of the "control" variable with the study variable and remove that portion of the study variable variance that is common with the control variable variance from the study comparisons. In the case of ANCOVA, this is generally accomplished by procedures that regress the study variable on the control variable, and adjust the study variable means by the application of the regression equation. The upshot of this sometimes rather complex procedure is that the study variable means may then be compared free of the linear effect of the control variable. Nonlinear cases become too complex for our consideration.

An example of such "correlative" analysis might be a study of two methods of teaching algebra. If we give a test at the end of the instructional period, we might wish to compare the mean score of students taught by Method I to the mean score of those taught by Method II. However, to make such a comparison meaningful, we might decide that any differences between the two classes with respect to intelligence test scores would overwhelm the effects of the two teaching methods, and so, we should "control for intelligence."

Without going into the pros and cons of this design or the computational aspects of this example, analysis of covariance could be used to remove the impact of differences between the two classes with respect to intelligence, thus permitting us to compare mean scores on the algebra test for the Method I and Method II classes, with the *effects of intelligence held out of the comparison*. The analysis of covariance is discussed further in the next chapter.

The computational aspects of partial correlation and covariance analysis are somewhat involved, and will not be treated here. However, we might say, in passing, that somewhat the same results can be achieved, crudely, by the use of procedures that might be termed "categorical covariance techniques."

In effect, the approach here is somewhat similar to the way that we approached the problem of curvilinear relationships in the previous section. In that case, we were unwilling to deal with the mathematical requirements for fitting curves to the points. As an alternative, we divided the predictor variable into categories and treated the criterion variable (got its mean) separately *within each of those categories*. Then we connected the category means in order to approximate a "line of best fit" for the prediction of *y* from *x*.

The analogous procedure here, categorical covariance, is simply to divide the data into categories based on the control variable, and then to perform the analysis of interest *within* each such category.

For example, in the instance referred to above, we might first divide the Method I and Method II groups into "High," "Medium," and "Low" subgroups on the basis of their intelligence test scores. Then we would compare the Method I High group algebra test mean to that for the Method II High group. Similarly, we would compare Methods I and II within the Medium group, and finally within the Low group. Of course in stating conclusions, particularly if they differed from intelligence level to intelligence level, we would have to be very careful to note to which subgroup each conclusion applied.

The approach of categorical covariance is only an approximation to that of partial correlation and analysis of covariance from the mathematical/

statistical model. It accomplishes the same goal, though less precisely and efficiently, but is certainly easier for many of us less mathematical types to understand and use. At the very least, it provides a way to think about more complex relationships.

The more precise, mathematical approaches can easily deal with the problem of controlling on several variables at once. Categorical control on more than one or two variables becomes too unwieldy and is unsatisfactory. On the other hand, *any* procedure attempting to control on more than a few variables tends to be somewhat unsatisfactory, whether or not the mathematics is tractable. This is because in the use of such procedures there is an increasing risk of basing results on increasingly large proportions of chance errors and fluctuations in the data as more and more variables enter into the computations. (We will have more to say about this point further on.)

IV. Contingency Analysis

Frequently the data from a study exist only in categorical form, and the question arises as to whether being classified in particular categories on one dimension is related to the classification into categories on the other. A simple two-way table is used to array the cases by category. It is called a "Contingency Table."

Suppose our two dimensions are "Socioeconomic Status" (SES) and "Voted in the last general election." The first has been categorized into High, Middle, and Low; the second has the categories of Yes or No. Suppose further that we have collected data on 50 persons with respect to these two dimensions. We might tally our results according to a table such as that shown in Table 7-1.

Table 7-1. Contingency Table for Chi-square Analysis

SES	Yes	Voted No	Total
High	8 (8.4)	7 (6.6)	15
Middle	15 (11.2)	5 (8.8)	20
Low	5 (8.4)	10 (6.6)	15
Total	28	22	50

You will notice that I have simply tallied each of the 50 people involved into a cell in the table representing his or her joint behavior on the two variables in question—and, that I have then totalled these tallies in each direction. The cell entries are called observed frequencies ("f"), and the row totals and the column totals are called "Marginal Frequencies," or "Marginal Totals." (I will explain the values in parentheses below.)

The question of course is whether or not this data table gives one a basis for supposing that voter behavior is related to SES status.

Any such table, laying out the categorical distribution of frequency of occurrence data on two variables jointly, is called a "Contingency Table." It makes no difference how many categories are employed on the two dimensions. There is a statistic that can be computed from such a display that can be used to ascertain the probability that such an arrangement is more than just the effects of chance. It is called the "chi-square" statistic (pronounced as in "eye" preceded by the "k" sound); its symbol is χ^2.

The chi-square statistic asks the question, "How much does the observed distribution of frequencies in these cells differ from that which would be expected if they were distributed proportionately to their marginal totals?" For example, there are 15 cases in the first row. If those 15 cases were to break out according to the distribution in the *total row*, there ought to be 28/50ths of the 15 in the first cell and 22/50ths in the second. Thus the "expected frequencies" (symbolically, "F") in the first row would be 8.4 and 6.6, respectively, and I have shown these values in parentheses beneath the actual frequencies.

It doesn't make any difference whether you do this by row or by column. For example, there are 28 frequencies in the first column. Using the *row totals* to apportion them, there ought to be 15/50ths in the first cell, 20/50ths in the second, and 15/50ths in the third, or 8.4, 11.2, and 8.4, respectively. I have entered the "expected frequencies" for all 6 cells in the table in parentheses for comparison with the corresponding obtained frequencies.

Now, as you might suspect, if the expected and the observed frequencies coincide, then we have observed exactly what was expected and the categorization on one variable is independent of the categorization on the other. That is, the number of observed frequencies in any cell is just a reflection of the split on the marginal totals, not a function of the distribution of the frequencies to the cells *within* the table.

But, to the extent that the observed frequencies differ from those that were expected, the arrangement with respect to the other variable made a difference beyond a simple proportioning according to marginal frequencies. Thus, it should be no surprise to find that the chi-square statistic is derived from these differences.

It should also be no surprise to find out that it is based on the square of these differences, because again minus signs can appear (the difference can be either positive or negative). In fact, since the expected frequencies sum to the same total as do the observed frequencies, if we tried to sum the differences it would always be 0 $[\Sigma(f - F) = \Sigma f - \Sigma F = 0]$.

It is tempting to try to treat the matter of an index of contingency by simply taking the mean squared deviation—as we did for distributions in general (i.e., the variance). However, there are two crucial differences in the situation. First, our data are not scores on a dimensional scale, they are simple counts; second, they do not range over a full dimensional scale, but are constrained twice in that they must add up to a fixed total on each dimension.

One of the difficulties in dealing with the squared difference is that its size is a function of the number of observations as well as whether or not there is any relationship between the two dimensions. For example, suppose the observed frequency were 5 and the expected frequency were 10; the squared difference would be 25. But, if the observed frequency were 105, and the expected frequency were 100, the squared difference would still be 25. It is intuitively clear that the first 25 is much more likely to represent a significant difference (relationship) than is the latter. Thus the larger the expected frequency in a cell, the larger the difference must be to suggest a relationship.

If, to neutralize the impact of size of expected frequency, we express the squared difference as a proportion of the expected frequency, we have then, in a way, standardized our scale of comparisons across the size of the cells. So, in the former example, the first 25 would be divided by 10, giving 2.5, and the second 25 would be divided by 100, giving .25. Doing this makes it clear that the first 25 suggests a much stronger relationship than does the second.

The procedure actually followed is to:

1. Take the difference between the observed and the expected frequency in a cell (this difference is the fundamental index of relationship according to the logic described above).
2. Square it (to eliminate the minus signs).
3. Divide the squared difference by the expected frequency (this in effect standardizes the cell index to the distribution of expected frequencies).
4. Repeat steps 1, 2, and 3 for each of the cells in the table (thus producing a set of cell indices of relationship).
5. Add the individual cell indices together to form a composite index (called chi-square).

In symbolic form:

$$\chi^2 = \sum (f - F)^2/F, \text{ taken over all cells.}$$

Chi-square is applicable to situations where the question is one of independence between categorizations on one variable and the same items categorized on another. Generally, the procedure works better if there are a limited number of categories among which to divide the number of cases studied. For example, it would be unlikely to find relationships in a system where the contingency table was 25 by 25. In general, most measurement situations don't yield 25 meaningful categories at any rate. Also, by then you are dealing with an approximation to a continuous variable, which may call for other methods.

Another point is that the system does not accommodate well cells with no observations in them. To avoid this, it is suggested that such categories be combined with adjacent categories, so that the final contingency table is of manageable size and is comprised of cells with solid numbers of observations.

Chi-square is a well-known and frequently used statistic. You should note that *it is always a positive number*—this is because the squaring takes care of any negatives in the numerators, and the denominators are always positive numbers. Thus, you must look to the distribution of the frequencies in the table for any guidance as to the direction of the relationship, if there is one.

If there is one! How do you know how large a chi-square must be to indicate a real relationship? We have not discussed this matter in our dealings with any of the relational measures presented. In the next chapter we will see how this question is approached. But, in the meantime, I will give you a crude procedure for converting the chi-square statistic into a measure roughly approximating r, so that at least you can have some sense of the degree of relationship contained in your contingency table.

The same Karl Pearson whose name is attached to the standard correlation coefficient suggested a statistic called the coefficient of contingency, based upon the chi-square statistic. It is obtained from a chi-square by taking *the square root of the fraction:* chi-square *divided by* chi-square *plus N*. (If interested in the historical background, see, Pearson, Karl. "On the Correlation of Characters Not Quantitatively Measurable." *Philosophical Transactions*, Series A, Vol. 195, (1901), pp. 1–47.) This coefficient is always positive (we noted earlier that in the case of contingency tables you have to examine the table to determine direction of relationship). In addition, it cannot reach +1. Because of these limitations it is not widely used.

V. Relationships Among More Than Two Variables

You have probably been wondering whether or not it is possible to deal with more than just two variables in problems involving relationships. Yes, of course. However, this gets into an area known as multivariate analysis, which is largely beyond the purposes of this book. Therefore, I shall limit my discussion of this topic to the single instance of using two or more variables together to predict one variable. This will give us plenty to think about, and should give you some idea of the basic concepts involved.

[Some of the topics (by no means all) that fall into this area, and which I omit, include canonical correlation, discriminant analysis, path analysis, meta analysis, factor analysis, logit and log-linear models, etc. In each case, both the logic and the mathematics of the topic become too complex for a treatise intended for the consumer of statistics rather than the career statistician. The reader is referred to the standard statistical references for information on these and related topics.]

Let us pause and review for a moment the topic of simple linear regression, as discussed in the previous chapter. You will remember that in that case we had one predictor variable and one criterion variable. We dealt with a graphic representation of these two variables in two dimensions (the x or predictor dimension and the y or criterion dimension).

We started with measures of x and y which were used to establish the predictive relationship upon which predictions of unknown ys from known xs were then based. The original measures of x and y were plotted on a scattergram, and any predicted y was given by reference to a regression line.

The regression line had a simple equation comprised of a slope that was related to the degree and direction of relationship and a constant that determined where the line of given slope crossed the y-axis. All predicted values of y lay along this line, and the measure of goodness of prediction was based upon the analysis of the amounts by which the obtained ys differed from the predicted ones. The general equation was:

$$\tilde{y} = bx + a$$

A. The Graphic Form of the Two Predictor Case

In multiple linear regression, we start out with exactly the same type of problem and the same logic. The difference is that although there is still only one criterion variable, there is more than one predictor.

Suppose you have two or more variables that you think may be related to a variable which, for whatever reason, you wish to predict. An example might be where students take a reading test and a math test, and you wish to predict their grade point averages at the end of the coming semester.

In such a case, we would determine the relationship from the test scores last year taken against the grades for the same students this year. We would then assume that the relationship was relatively constant over time, administer the tests to new students, and use their test scores and the relationship to predict their grades next year. In this example the reading test score and the math test score are the "predictor" variables, and students' grade point averages are called the "criterion variable."

For simplicity, we will discuss the two-predictor case, as in our example above, understanding that for larger numbers of predictors the logic is identical, although the mathematics rapidly become formidable. Symbolically, we will use x_1 and x_2 to represent the two predictor variables, and y to stand for the criterion variable. So, for each of the cases in our study we have *three* measures instead of two: two predictors and one criterion variable, or x_1, x_2, and y.

In the case of multiple regression, even with only two predictors, it is not convenient to produce a scatterplot. This is because we now must plot the points in *three* dimensions, whereas before we had only two. There must be a dimension for the x_1 value to be measured along, another for the x_2, and a third for the y. (You can think of this as like the dimensions of a room, with the predictor variables being length and width on the floor, and the criterion variable being the height dimension.)

If we plotted the original grades along with the predictor variables, math and reading scores, we would get a swarm of points in three-space. Each of the points would have its own three coordinates (location) determined by measuring out along each of the three dimensions involved to the proper values. In the example of the room, the points would be hanging at the appropriate height in midair, each over its corresponding x_1, x_2 combination on the floor.

(Now you can see why I wanted to restrict our discussion to the two-predictor case of multiple regression. Each additional predictor represents an additional dimension! Three predictors requires a 4-dimensional problem, and so on. Mathematicians may be used to thinking in what they call "hyperspace," but most of us would rather not. By sticking to the two-case we at least stay in three dimensions, with which we have all had experience.)

Now, let's consider the predictors. Theoretically any combination of values of x_1 and x_2 is possible (each combination of which has a predicted *y* associated with it). How are these combinations distributed? They certainly don't lie in a straight line! In fact they could be anywhere on the "floor" of our three dimensional room, within the constraints of the range of measurement on the two variables.

Thus, the *y* values associated with these combinations could also be anywhere depending on their "length and breadth" measurements. In addition, the *y*s also vary up and down along their own dimension.

This is important because, of course, our goal is to find some entity of "best fit," based on our predictor variables, that best estimates the real *y* values. In the simple linear regression case, that was a straight line, but as I have noted, a straight line won't do here.

What I am trying to say is that the swarm of points is not distributed so that a simple line of best fit can be passed through them. This is the result of the fact that the two predictors taken together form a *two*-dimensional system whereas a line is only a *one*-dimensional entity. The surface of best fit for a system where the predictors are a two-dimensional system must itself be a two-dimensional surface.

In fact, the prediction surface is a plane instead of a line. [Think of a plane as being like a pane (of glass), having length and breadth, but no thickness.] A plane is defined (fixed) in any one of several ways: by three points not in a straight line; by two intersecting straight lines, or by one straight line and a slope angle between the plane and the plane of the predictor axes.]

Figure 7-2 shows a schematic illustration for predicting grades from reading and math. The prediction plane is shown, as well as both the predicted and actual grades for a set of 10 cases.

So, where in the linear case we sought a straight line constructed so as to minimize the amounts by which it missed the actual *y* values, here we seek to pass a plane through the swarm of points in three-space so that the distances of the actual *y* values from the plane (which consists of the predicted *y* values) are minimized overall. [Actually, as in the linear case, the mathematical requirement is to minimize the *sum of squares* of the deviations between the actual (y) values and the predicted $\tilde{y}$ values represented by the predictor plane.]

Now, if we had no knowledge about *y*, as a function of the various combinations of x_1 and x_2, our best prediction would be, as always, the mean of the *y* variable. That would result in each predictor combination, no matter what (or where on the "floor") having the same *y* value "above" it. In effect those points would define a plane parallel to the "floor." Or, in other words, under those conditions, the prediction surface would be a plane above the predictor plane and separated from it by a

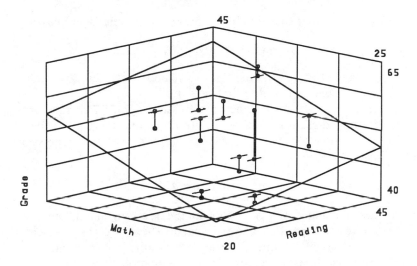

45

25

65

40

45

Grade

Math

20

Reading

—•— = predicted values and lie on the regression plane

• = actual values

Figure 7-2. Three-Dimensional Scatterplot

distance equal to the *y* mean—that is, parallel to the plane of the two predictors.

Of course, if the predictors provide knowledge about the criterion variable, as would normally be the case, then we would find differing values of *y* depending upon the differing values of the combinations of the predictors. Thus, in this case, the prediction plane would not be parallel to the predictor plane, but rather would have a relationship to it determined by the interrelationships of the three variables. Figure 7-2 illustrates such a predictor plane for the data in this small artificial set.

B. An Equation for the Prediction Plane

How do we know that the predicted values lie in a plane? This is easy to visualize when each predicted value is the *y* mean, but not so easy in the more likely case where the predictors are functional. I will not offer a mathematical proof here, but simply state that our mathematician friends have demonstrated that "Every plane is represented by a linear equation (one in which no exponent of any variable is higher than 1) in three variables, and, conversely, every such linear equation represents a plane."

This situation means that the general equation for a plane is as follows:

$$ax + by + cz + d = 0,$$

where a, b, c, and d are constants, and x, y, and z are three variables, each with an exponent of 1 (to the first power).

The general form of the two-predictor regression equation fits this form, once it is rearranged slightly. (You will note the similarity to the general form of the equation for one predictor.) The two-predictor regression equation is as follows:

$$\tilde{y} = b_1 x_1 + b_2 x_2 + a,$$

where we have combined the constants in the rearrangement, but note that we still have the three first power variables (here, the two predictors and the criterion variable). (Note also that if we subtract $\tilde{y}$ from each side of this equation, we have, as in the previous equation, three first-power variable terms and a constant term all equal to zero—just as in the general equation for a plane.)

(As a matter of information, the general equation for "n" predictors is quite similar, simply adding another term for each additional variable: $+b_3 x_3 + b_4 x_4 \ldots + b_n x_n$. However, the addition of more variables makes the computation of the constants, the bs and the a, correspondingly more difficult, because these values depend on the means, SDs, and intercorrelations of all the variables in the problem.)

What do we know about the plane we are looking for? We know offhand that our required plane must contain two points: the point where the triple lines of means cross (because if the first predictor value is its mean, and the second is its mean, the best prediction of y is the y mean). Also, because the prediction plane must accommodate, at least theoretically, a predicted y where the predictors have the value of 0, there is a y-axis intercept value.

Thus, the plane must contain the line running from the intersection of means to the y-axis intercept. The other determinant of the predictor plane is its slope. This slope is a function of the interrelationships (correlations) of the variables involved, and is related to the strength of the predictive relationship, but in a more complex manner than in the case of linear regression.

As noted above, the general case for the two-predictor problem is as follows:

$$\tilde{y} = b_1 x_1 + b_2 x_2 + a,$$

where the bs, called regression coefficients, control the slope function and are derived from the correlations among the variables and (to equalize units) the standard deviations; and the a represents the y-intercept of the plane, and is a function of both the bs and the means of the variables.

If all of the variables are converted to standard distribution form, the a constant drops out, and the SDs are no longer required, so that the bs become strictly functions of the correlations among the three variables. In this case, the plane must pass through the common zero point, or origin, at a slope angle to the predictor plane determined by the intercorrelations.

The procedure in using a system such as this to predict y is analogous to going to the point on the "floor" represented by a particular x_1, x_2 combination, reading up to the plane, and then reading across horizontally to the y value corresponding to that point on the plane.

C. The Multiple Correlation Coefficient

We are of course interested in some index of the relationship present in any multiple regression problem. You can think about such an index in one of two ways. First, if you should apply the standard (Pearsonian) correlation techniques to the two-variable set comprised of the actual ys and the predicted ys, you would get a correlation coefficient that directly represents the degree of relationship between the predicted values and the obtained values. This value is usually called "R," the "multiple correlation."

The interpretation of R is similar to that of any other correlation. It tends to be high when the predicted values are close to the actual values of y; it tends to be low when the predicted values diverge from the actual values. Of course, (unattainable) perfect prediction would produce a value of 1.0. However, it is unlikely that R would ever be as small as zero, as one would always tend to pick predictors that have at least some relationship to (correlation with) the actual y values. In addition, negative values are precluded by the mathematical relationships underlying the calculations.

The R can also be interpreted through the use of the coefficient of determination (R^2). As with linear regression, R^2 represents the proportion of the variance in the criterion variable that is accounted for, or common with, or predicted from, the combination of the predictors, as optimized by the computational procedures.

The second way to think of an index of multiple relationships is to analyze the errors of prediction as we did before with linear correlation. It can be shown as before that the total variance of the criterion variable, V_y, can be broken into the predicted variance, $V_{\tilde{y}}$, plus the error variance, V_{error}. Thus, R may also be estimated from such an analysis. Finally, the analysis of the deviations of $\tilde{y}$ from y produces a standard error of estimate equal to the SD_y times the square root of $(1 - R^2)$, entirely analogously to the case of simple linear regression.

D. The Interpretation of "R"

Before concluding this section, there are several points to be raised about the use and characteristics of multiple correlation. First, as hinted above, it is possible to generalize the above relationships to many predictors. But you should understand that each additional predictor brings with it into the system additional errors of measurement. This is important, because the mathematical process of determining the surface of best fit to the points in hyperspace tends to capitalize on chance errors in the data. The process has no way of knowing whether a given score has its value because it is correct or because of some error of measurement.

The import of this is simply that random addition of predictors to a problem is ill-advised. It tends to inflate the estimate of the multiple correlation as these chance factors accumulate. Theoretically, it is possible to approach perfect *though spurious* prediction simply by including enough predictors in the equations, whether or not they have any logical role to play in the problem at hand. So, studies should be designed with solid conviction that the predictors are meaningful, rather than throwing in every variable in the book!

A second and related point is that this capitalization on error or chance fluctuation takes place even with well-chosen sets of variables. This is always true to some degree. About the only way to assess the extent of this effect is to repeat the study with fresh data, but using the old constants in the prediction equations, and examine the extent to which the R value shrinks (which it will, since a new set of chance factors has been introduced). This process is called "cross-validation."

In choosing predictors to enter into a multiple regression problem, the first issue of course is to choose them so that they have high correlations with the criterion variable. It seems indisputable that if the predictor is unrelated to the criterion, it cannot function as a predictor! But, it is also of concern to examine the extent to which the predictors are related to each other. The more closely two predictors are related to each other, the less likely that the two together will contribute much to the prediction beyond what either one would yield by itself. In other words, they are likely to overlap.

To take an example, variables a, b, and c are each related to y with a correlation of .6, and a and b are related .8, whereas a and c and b and c are related at .3. In this case, it is unlikely that all three variables will contribute much more to the prediction than using either a and c or b and c; and, we surely wouldn't use just a and b, which are highly similar.

Finally, it is possible, by looking at the relative size of the bs in the prediction equation, to ascertain which predictors in the problem are carrying the bulk of the weight in making the predictions. The larger the b, the more of the predictive weight it is carrying in this particular problem. However, this is tricky. If you do another problem, including a given predictor variable with a different set of other predictors, you frequently find that the b value has changed considerably in this new setting. In other words, the bs are relatively unstable across different mixes of predictors.

The reason for this is of course the interrelationships among the predictors. A predictor that was strong in one instance, if included in a mix with another predictor to which it is highly related but which is a little closer to the criterion, may see its b value drop drastically, as the new predictor takes over the weight the first one previously had.

To sum up, the contribution of any predictor is a function both of its relationship to the criterion variable and its intercorrelations with any other predictors in the problem. If two predictors are unrelated to each other and the other predictors in a problem, their predictive value becomes strictly a function of their respective correlations with the criterion variable.

If they are moderately related to each other, their total predictive value tends to be greater together than either would have alone. If they are closely related to each other, then the use of both instead of one does not improve prediction, and indeed, may provide a falsely inflated multiple correlation (due to the tendency of the process to capitalize on the additional errors added to the mix by the unnecessary predictor).

Another caution to be observed is that the use of the same predictor equations over extended periods of time is likely to be unwarranted if the variables involved are subject to change across time—as many are. Again, if the population on which the equations were determined should differ significantly from that to which the prediction is applied, the predictions won't hold. Or, if either of these populations changes significantly across time, later applications will be unwarranted.

As a result of these and other limitations inherent in this process, the prediction equations should be checked periodically by analyzing predicted values against actual, obtained values to assess the goodness of prediction on an ongoing basis.

VI. Some Reflections on Correlation

Although I'm sure you will agree with me that relationships are important, we have spent so much time in the last two chapters developing the fundamental ideas underlying the measurement of relationships that we may have lost sight of some of the larger questions, such as: When I see a correlation in a report, what does it mean? What are some of the guidelines for interpreting these things? Where might I expect to find them? What factors affect correlation coefficients? How widely applicable is the process of correlating variables? Is there such a thing as relationships without correlation coefficients? Etc.

I don't think that I can answer all of these questions to your satisfaction, but perhaps we can develop a sense for correlations. In the comments to follow, I will talk about "correlation" as though I were dealing only with the usual product-moment (Pearsonian) type, but the comments are largely applicable to all of the varieties we have discussed above.

A. Applicability

Calculating a correlation coefficient is a straightforward mechanical process. It can be done for any set of paired numbers. Whether the result is meaningful or not is another issue! The first question you should always ask is not directed at the numerical results of a study, but rather at the soundness of the logic underlying the study. Does the problem studied make sense? Do the measures used fairly represent the concepts involved? Is the design fair, or is someone trying to prove his own point? Has the data collection been conducted competently? Is the analysis sensitive to the issues in question?, and so on.

Given satisfactory answers to all of the above, correlations are descriptive measures. They merely assess the strength and direction of the tendency of two sets of numbers to vary jointly. They are insensitive entirely to whatever it is that those numbers may represent. They are insensitive to whether or not the relationship measured is causal. They are, however, or at least their interpretation is, sensitive to certain assumptions about the distribution of the scores or numbers from which they are calculated.

I shall not go into detail here about the assumptions, which apply largely to certain tests that statisticians apply to ascertain whether or not the relationship represented by the correlation coefficient is strong enough to make any difference. I do, however, want to mention a factor that will directly affect the size of the coefficient itself. It is the assumption that each of the variables being correlated is free to vary without artificial restriction, in other words, the assumption of "unrestricted range."

B. Restriction of Range

To use the "reductio ad absurdum" approach again, suppose you selected a group of boys to do the height and weight study in the last chapter, but you correlated height and weight for a group of boys all of whom were 66 inches tall. Perhaps you would be surprised to find that the correlation coefficient would be zero! But consider—the correlation is a measure of *co*-variance. If one or both of the variables don't vary, you can't have any co-variation!

Pooh, you say. Anybody can see that. Yes, but this effect occurs in lesser degree in many cases where it is not so easy to spot. Suppose you do our height and weight correlation for a physical education class of boys *who are all candidates for the school basketball team*. You will find a smaller correlation between height and weight than you would in the general population of boys of similar age.

Why? Because basketball players are all likely to be considerably taller than average. As compared to the general population, you will have restricted the range, and thus reduced the variance, on the height dimension. The inevitable effect of having less height variation is to reduce the height co-variation (correlation) with *any* other variable.

The effect of restriction of range can crop up in many places. The correlation of aptitude test scores with high school grades is less for college students than it is for all high school graduates. This is because a select group of high school graduates, aptitudewise, went on to college, whereas less able graduates did not.

If we are only interested in studying college students, this correlation may be sufficient for our purposes. It does describe the situation in the population of interest. However, if we are interested in studying high school graduates, we must realize that the correlation is "attenuated" (shrunken) by restriction of range. Thus, the import of this phenomenon is relative, depending on the focus of the study.

I should note in passing that it is also possible to use the range phenomenon to raise correlation coefficients! How can this be done? If I ring in some cases that normally wouldn't fit the definition of the study, but who represent extreme cases on one of the variables being correlated, the existence of this "excess" range will operate to increase the co-variation and raise the correlation coefficient.

This is one reason that it is unwise to include excessively broad groupings of subjects in a study—such as correlating reading scores and grades over a several grade grouping. Teaching techniques are oriented to within grade procedures, and single grade correlations would be more defensible.

C. Significance

We statisticians speak of something called "significance." I will talk some about what that means in the next chapter. But, for now, it means that a result, such as a correlation coefficient, is big enough so that it probably represent an actual relationship, as opposed to one that arose from chance factors or coincidence.

Beware of reports that offer tables of very small correlations that are described as "significant." These often arise where large numbers of observations are correlated. They result from the fact that the statistical procedure used to establish significance gets more and more sensitive as a function of the number of observations included in the determination. Therefore, it is only necessary to collect enough cases in order to make almost any size correlation "significant" by statistical standards.

This is where the coefficient of determination (r^2) is a lifesaver. Even if a correlation meets some criterion of statistical "significance," it is still legitimate to ask, "But does it make any practical difference?"

For example, if we find out that the correlation between painting classroom walls gray and student's grades is, say, .10, are we justified in undertaking a massive repainting job on all of the classrooms in town? Maybe, maybe not. Squaring .10 gives us .01, which means that 1% of the variance in student grades is associated with gray paint. Even if we were to assume that this is a causal relationship, rather than just associational, the magnitude of real impact a one percent change would make might not justify the major expense of repainting all of the school rooms in town.

D. Some Comments on Causality

I will close this chapter on relationships by turning to the area of causality. Causal relationships are of course what science is all about. It is useful sometimes to know that things are related—tend to occur together or similarly—but knowing that something causes something else implies all sorts of possibilities for taking controlling action. As a result, people have a very strong inclination to attribute *causation* to situations even though it may only have been demonstrated that a *relationship* exists between two variables. Frequently this is wrong, and even leads to mischief.

The impact of education on income is such an area. In the early 1900s the enthusiasm for education as a means of personal betterment was such that the U.S. Office of Education itself put out posters describing the worth in dollars of obtaining a high school education. It based these on studies showing that the average high school graduate earned hundreds of thousands of dollars more over a working lifetime than did non-high-school graduates.

That finding was correct, and, indeed, it can still be demonstrated today that, on average, high school graduates earn more than non-high-school graduates, and, in turn, that college graduates earn more than either. But the question is whether the increase in earnings is directly attributable to the additional education. Is it true that all we have to do is to keep children in school longer? To some extent our compulsory education laws seem to be based on this philosophy.

Although clearly there is a correlation between education and earnings, the causality of the situation is far more murky. To continue with the example, suppose it was the smarter students who stayed in school and got high school educations; but, smarter people also tend to be more financially successful. Was the increased earning power of high school graduates, as observed by the U.S. Office of Education, a function of their greater education, or of their greater intelligence?

Similarly, it was said of one of the graduating classes of Yale (somewhere around 1910, I think) that more than half of them had their own businesses or were corporate officers within 10 years—such is the power of a good college education. But note that a high proportion of these graduates had gone to work in family businesses or in positions in companies where their fathers had senior positions. Was their success a tribute to education at Yale—or to family connections and "pull"?

As a matter of fact, as we have become more sophisticated in designing studies and using statistics, we have recognized that correlational relationships rarely have a simple and obvious connection with causation. For example, to demonstrate the value of education for occupational success, it would be necessary to conduct an experiment something like the following:

Take a large group of identical twins reared together (to control both heredity and environment); then separate them, sending one set to college and the other set out to the job market; finally, compare their successes some years later. Only in this way (by holding hereditary traits and upbringing constant) could we get an unbiased idea of the extent to which the educational process did or did not contribute more to occupational success than did the additional experience gained by starting one's career four years earlier.

Let me digress a moment to add a few comments on experimental design at this point. The above example is not intended to be a full experimental design, but merely to illustrate one difficulty in this area.

There are those who would say that the assignment of the identical twins to the two groups (college and non-college) must be done on a randomized basis in order to assess the causality of the education relationship. I don't see the need for this in this instance, though it can cause no harm. In addition, there are many situations where such a procedure is not only desirable, but essential. Basically, as noted in Chapter 2, the real issue is always control of unwanted and/or unknown influences on the results of the experiment.

Randomization is a control technique whereby we depend on the action of chance to distribute unknown or otherwise uncontrollable factors equitably among the study groups. Randomization does *not* however *guarantee* that causality can be established. Indeed, blind reliance on randomization can be misleading, as chance can also result in serious misdistribution or concentration of unwanted influences in one or another of the study groups. Chance is, after all, just chance!

In the identical twins experiment described above it seems unlikely that there are important factors that remain to be controlled, and that randomization at the point of high school graduation could effectively control. Each twin would have had identical heredity and highly similar upbringing up to that point. Randomization wouldn't hurt, and, given that doing it entailed no extra costs or inconveniences, should probably be done. But, the lack of randomization does not significantly impair whatever implications of causality obtain in the experiment as described.

By far the more difficult question to me would be the problem of defining what is meant by occupational success, which is a much larger and more complex concept that just earned income.

To return to the twin situation, needless to say no one has run such an experiment, so that we are still not sure of the *real economic* value of education *to the individual*. (This is not to impugn the value of education from a cultural and social standpoint, or to our nation as a whole, of course. It is not my purpose to suggest that it is of no value economically, only to suggest that to establish a causal relationship between education and income or occupational success is much more complex than you may have thought!)

Let me cement my point with two more examples. First, consider that dying and being in bed when you die are very highly correlated. Do we conclude from this relationship that going to bed causes you to die? Finally, consider the fact that the correlation between the birthrate in one Scandinavian capital and the number of stork's nests in the city over the years 1930–1950 was a very respectable .76. Do we take this as evidence in support of the old theory that the storks are bringing the babies...?

In fact, there are four conditions that can give rise to a significant correlation between variables A and B:

1. A causes B.
2. B causes A. (The value of the correlation is the same as for #1, but the implications are vastly different for a study and its conclusions.)
3. A and B are both caused by some other variable(s), known or unknown.
4. A and B are actually unrelated, but chance factors caused the apparent correlation (particularly since the correlation process inherently tends to capitalize on chance fluctuations in the data, so that chance factors can combine to give the appearance of real relationships.)

In a real study, you need to sort out the logic of these four possibilities yourself, based upon your understanding of the issues! In our height vs. weight example, it would appear likely that the relationship that we observed in our data would be an example of possibility #3. The other factors would probably have to do with genes and heredity, as well as learned eating behaviors, etc.

In the case of dying, we have obviously confused the issue by failing to sort out the reason for going to bed, as well as imputing cause in the wrong direction (possibility #2). As for the stork's nests, without further data it is impossible to say whether this is possibility #4 (which can sometimes be described as a "coincidence," or possibility #3 where we don't have the information to identify the true causality.

I have tried to make the point above that many times correlational relationships are not causal, and that we must guard against the unconfirmed assumption that a relationship implies causality. Unfortunately the difficulty in confirming causality extends to the other side of the coin as well. That is, there are sometimes cases where the correlational relationship *does* represent causality, at least to some degree, but this is also, for one or another reasons, difficult to confirm. A famous case in point is the relationship between cigarette smoking and lung cancer.

Relatively few people disagree any more that there is a causal relationship (certainly not a perfect one of course) between cigarette smoking and contracting lung cancer. The few remaining skeptics stoutly maintain that causality has not been proved. They're right. As is frequently the case, the appropriate scientific experiments are too dangerous or unethical to conduct. Yet, over a period of years, most doubts about this relationship have been laid to rest. How did we manage to get from a correlational relationship to a causal one in this matter?

E. The Quasi-Causal Net

I have termed this process "constructing a 'quasi-causal net' (QCN)." The concept of a QCN recognizes that some relationships, although they are relatively easy to establish as relationships, are difficult or impossible to prove causal. The philosophy is simply to establish and reestablish the correlational relationship again and again under differing circumstances and with differing groups. Gradually, over a period of time the variables competing for a causal role get eliminated or become less and less likely, and correspondingly the correlative variable becomes more and more probable as a causal factor.

Obviously, a QCN can never approach certainty, but may, as in the case of smoking and lung cancer, become sufficiently indicative of probable cause that constructive actions may be considered.

To go back to the education arena, to establish the QCN for education as a cause of improved earnings, you would run the correlation for a variety of groups in a variety of settings. First, you would try it for men and for women; then by age groups, then by occupational groups, then by various levels of education, then within differing socioeconomic strata and geographic locations, then perhaps in differing cultures, etc.

In so doing, given that the relationship continued to occur under all of these various situations, you in effect eliminate age, sex, occupation, culture, etc., by showing that the relationship holds in spite of differences on these variables. Thus, slowly you weave a net of relationships around the causal principle, strand by strand.

Obviously such a systematic sequence of studies is an expensive approach, and generally not warranted unless the issues concerned are of vast range and consequence, and not approachable by other means. It is equally clear, however, that we must systematically build QCNs where great human consequences depend on the establishment of causality, but controlled experimentation is limited because of its ethical, moral, or physical consequences.

VII. Summary

We began the consideration of relationships with variance and covariance in the previous chapter and worked our way through the topic of relationships and simple linear regression. In the present chapter, we found that there are several variations on simple linear regression and correlation, designed to fit special situations such as the use of ranks, dichotomous variables, nonlinear situations, and categorical cross-classifications. All such procedures are somewhat similar to the basic correlation procedures, both in intent and in interpretation.

I mostly avoided multivariate analysis as being too complex for our purposes here. However, I did spend some time on the simplest case—that of the two-predictor problem. I showed that adding a second predictor to the simple linear case in effect adds a dimension (and that additional predictors would add additional dimensions). Thus, in the two-predictor case, instead of looking for a line of best fit between criterion and one predictor, we are looking for a plane of best fit between criterion and two predictors, and all predicted ys will lie on that plane. The general equation for the plane was given.

We examined an index of the degree of relationship represented by multiple regression problems. The multiple correlation (R) has meaning and interpretation rather similar to that of the standard Pearsonian coefficient. It can be defined by applying the Pearsonian process to the correlation of the actual criterion scores with their predicted values. The coefficient of determination (R^2) is similar in meaning to its Pearsonian counterpart.

Some cautions were offered with respect to the use of multiple regression, most notably the fact that the process capitalizes on error in the data, and that predictors should be selected with care, both in number and nature. It was suggested that all such studies should be subjected to "cross-validation" to assess the extent of inflation in the results as a function of error.

The chapter concluded with an outline of some of the constraints on the use of correlational techniques. Mentioned were the attenuations caused by restricting the range of one or both of the variables; the importance of appropriateness of application; the ease with which small correlations can be made significant, and most importantly the pitfalls associated with an unwarranted assumption of causality. Simple relationship (correlation) should not be mistaken for causality. Variable A may cause B—or vice versa; both may be caused by C; or the observed relationship may simply be chance or coincidence.

It was noted that many times causality is assumed under circumstances where it is exceedingly difficult, if not impossible, to confirm it. This is not just a matter of ignorance or willingness to assume causality. Very often the critical experiments or tests do not lie within the bounds of economics or ethics or morality. A case in point is the relationship between smoking and lung cancer. In such cases it is necessary to go through the long-drawn-out and tedious procedure of building a quasi-causal net—increasing the probability of causality through a process of reducing the probability of competing factors.

VIII. References

Chatfield, Christopher and Collins, A. J. *Introduction to Multivariate Analysis*. New York: Chapman and Hall, 1980.

Cliff, Norman. *Analyzing Multivariate Data*. San Diego: Harcourt Brace Jovanovich, 1987.

Dunteman, George H. *Introduction to Multivariate Analysis*. Beverly Hills: Sage Publications, 1984.

Everitt, B. S. *The Analysis of Contingency Tables*. London: Chapman and Hall, 1977.

Fox, John. *Linear Statistical Models & Related Methodology*. New York: Wiley, 1984.

Gordon, A. D. *Classification Methods for the Exploratory Analysis of Multivariate Data*. New York: Chapman & Hall, 1981.

Jaccard, James, Turrisi, Robert and Wan, C. K. *Interactive Effects in Multiple Regression*. Newbury Park, CA: Sage, 1990.

8

The Nature of Statistical Inference

I. Overview and Purpose

In the previous several chapters we have been concerned with the problem of eliciting information from a body of data. We talked about what

243

distributions can tell us, what central tendency can tell us, the meaning of variation, and, finally, co-relationships of various kinds. Taken together, such topics are termed "descriptive" statistics, because they enable us to describe the information contained in the data we have collected.

They do not, however, by themselves, enable us to educe general conclusions or state principles—they merely enable us to describe what is there in the data at hand.

In order to go beyond simple description, we must turn to the second major branch of statistical technique, "inferential" statistics. Inferential questions exist whenever we wish to make general statements from partial information.

For example, if we are able to collect information about only a part of a population, but wish to make a statement about the whole population, we must "infer" the general condition from the specific part we have studied.

Or, we may wish to examine the difference in outcome between two or more "treatments" or procedures that we have applied. If we observe such a difference, we must again "infer" that the difference would apply to the entire population if it were exposed to the same treatment conditions; or, perhaps, that it would be seen again with the same subjects if the treatments were to be repeated at a later time or under somewhat different conditions. Such questions are among the essential aspects of the use of the scientific method to advance human knowledge.

In this chapter, then, we will examine how statisticians go about the task of addressing inferential questions. We will look first at the logical aspects of such endeavors, then at the major mechanisms that are employed to attack such questions. Finally, we will consider some of the specific techniques used, and their applications. In all cases, however, we maintain our focus on the principles involved as opposed to the mathematics and the calculations, because these techniques can be just as technically complex as the area of multivariate analysis discussed in the preceding chapter.

II. The Nature of Statistical Inference

As we have seen in previous chapters, there is much that the statistician can contribute to the understanding of the information contained in the data collected in various endeavors. But the statistician cannot stop with the application of various descriptive and relational techniques. There are other questions that are frequently of even more interest to be examined. These are questions about whether or not the findings from a set of data hold for other groups, or for future groups, etc. These are questions of inference.

A. *Partial Information and Inference*

In general, inference is the process of generalizing or drawing conclusions based on less than complete information. In statistical inference the reason that we have less than complete information is typically that we have not looked at the entire universe of objects, items, or individuals under study.

Thus, it follows that statistical inference techniques would not be necessary if we could simply examine the entire universe of interest with descriptive techniques! (This, of course, does not mean that looking at the entire universe guarantees complete accuracy—for that, perfect (error-free) measurement would also be needed.)

It is not always possible to examine the entire universe for many reasons: Not enough time, money, or manpower; the population may be too big to measure; the measurements may be too involved, difficult, or expensive; or, the measurement process may destroy the test objects. Then, too, the population may be unavailable for measurement, either physically, or because some part of it is future events. Etc.

Therefore, statistical inference always deals with partial information; thus, it involves making statements about a whole population, or about an unknown part of a population, based upon a known part. (I say always, for, in truth, I know of no nontrivial exceptions.)

This leads to another and exceedingly important point. That point is that you can never prove anything absolutely with statistical inference. Because part of the population is, and remains, unknown, even the most sophisticated estimate of the unknown part retains an element of doubt. How can you be sure about something that is unknown?

As an illustration, suppose I have developed a scientific hypothesis that I wish to prove. My hypothesis is that (for reasons that are irrelevant here) somewhere in the world there is a purple, polka-dotted bird. In order to test my hypothesis I travel the world searching for such a bird. Although I travel many thousands of miles and screen literally millions of birds, I do not see the bird I am looking for. What is my conclusion?

The fact is that I can only state that I have failed to find such a bird—*not* that there is no such bird! Not only have I not proved the existence of such a bird, I haven't conclusively demonstrated its nonexistence either. Unless I have looked at *every* bird in the world (in which case we are talking descriptive, not inferential, statistics), the bird I seek can still exist among those not yet examined.

Further, unless I actually see such a bird, it doesn't really make any difference whether the hypothesis was that there is such a bird, or that such a bird is impossible. The simple fact is that statistical inference, because it always involves generalizing from partial information, does not guarantee the test (proof) of *any* assertion. The possibility always exists that the critical information may reside in that portion of the universe not examined!

Because, obviously, we engage in statistical inference for a purpose, to make supportable statements about a population, our inability to prove a hypothesis with statistical inference is something of a liability. How do we deal with this difficulty? The answer is that statisticians have developed a logic for statistical inference which features "indirect proof."

The idea of indirect proof entails one more concept that I want to mention before I try to describe the process—that is *probability*. Probability was mentioned earlier when I told you in Chapter 2 that there are some pervasive ideas upon which statistics is based. I have demonstrated in the material on descriptive statistics that the concept of averages is one of those ideas; it certainly pervades descriptive statistics. The bedrock of inferential statistics is another one of these overarching ideas; it is the notion of probability. Because we can't prove anything with inferential statistics, we use indirect proof, bolstered by statements of probability. Thus, inferential statistics is intimately bound up with the idea of probability.

I will return to the matter of probability in inferential statistics later.

B. Statistical Logic

Of course, logic is logical. But statistical logic is something else again! It has a couple of fillips to it that grow out of the problem of trying to prove something that cannot be proved. Let's use an example to look more closely at why you can't prove something with inferential statistics.

We have shown that we cannot *prove* either side of a hypothesis, because of the partial nature of inferential data. How about *disproof* of a hypothesis? Yes, this is possible. All it takes is for the critical event to be included *in that part of the population we examine*. In our bird example above, if the hypothesis is that there are *no* purple, polka-dotted birds, and, in my examination of 1000 birds I do find one, then surely the hypothesis is disproved, or discredited!

We now have the key to the statistical logic of indirect proof. What is needed now is to frame the question as a set of *two* hypotheses, *which are mutually contradictory*. Then, if one is discredited, the other must be supported, indirectly.

In practice, the hypothesis of interest, or study hypothesis is stated. Then, a "contradictory" hypothesis (called the "null" hypothesis) is posed and tested. If the data provide a reason to reject the contradictory, null hypothesis, we consider the study hypothesis to have been "proved" indirectly (really "supported," since the rejection of the null is rarely absolute), and a probability statement of being wrong is attached.

Let me take a more reasonable example. Suppose I were a referee whose job it was to toss the coin at the beginning of football games, so that the kicking team could be determined. I would certainly want my coin to be unbiased (fair). However, I might suspect that my coin favored heads, and want to contrive a test to see if it were truly unbiased.

Before starting, I would set up my study hypothesis (H_s): "The proportion of heads exceeds 50%". Then I would set up the null hypothesis (H_n): "The proportion of heads is 50% or less." Actually, since I wouldn't want the coin to be biased in *either* direction, I would probably set the hypotheses up as follows:

$$H_s: \quad Hds > 50\% \text{ or } Hds < 50\%;$$

$$H_n: \quad Hds = 50\%.$$

Then I would test the null by tossing the coin, perhaps 200 times, and tabulate the results. If my data suggest that the coin's percentage of heads is other than 50%, I have discredited my null hypothesis, and have supported my suspicion that I had better get another coin before going back to my job. After all, those football players are big, and can be rough!

On the other hand, if the findings came out so close to 50% as to make 50% believable, my conclusion must be that "*I failed to reject the null hypothesis.*" This is all that can be said. You cannot say that you proved the null, only that you failed to reject it. This is because the results might have been different if you had taken another 200 tosses. (Remember, no hypothesis can be proved directly.)

In summary, out of such complicated reasoning statisticians have established the procedure for testing their hypotheses: Because you can't establish any assertion directly, you set up a test of a hypothesis that is the exact opposite of, or contradictory to, the hypothesis that you really want to test. Then you look for evidence to discredit your second hypothesis. If found, such evidence is, indirectly, supportive of your original hypothesis. This process is known as setting up contradictory hypotheses—that is, hypotheses such that if one is true, the other must be false, and vice versa. Then when the one is discredited, the other is automatically supported.

The test hypothesis is called the "null" hypothesis. For example, if you contend that a treatment that you have employed, perhaps a new teaching method, has made a difference in students' reading scores, you test the (null) hypothesis of no difference. If this can be rejected, then you have support for the conclusion that the treatment did indeed make a difference.

But suppose that the test of the null hypothesis turns up no evidence to discredit it. In that case, you must accept the null hypothesis—at least for the time being! You *cannot* consider the null hypothesis to have been proved—statistical data is not capable of proving the null any more than it is of any other hypothesis.

In passing, I should note that it may still be possible to reject a null that you have accepted, if a more precise or more powerful test of it can be devised and fresh data collected. Whether or not the matter warrants such additional efforts depends upon the circumstances. (Graduate students, for example, sometimes think that their doctoral theses cannot stand upon a null finding, and have frequently been observed collecting more data and contriving new ways of testing their propositions!)

Although this procedure of testing the null hypothesis may seem cumbersome, it is logically necessary. In addition, it turns out that statisticians have been more successful devising appropriate tests of the null that they have of the real hypothesis.

C. Two Kinds of Error

Remember, the object of statistical inference is to make some kind of statement about a population some portion of which is unknown. Such a statement might be about a mean, or a SD, or a difference in means, or some such. As you may have inferred from the above discussion, regardless of what the statement is, there are two ways to go wrong:

1. You can say something is so, when it isn't.
2. You can say something isn't so, when it is!

Logically, each of these two possible errors could be important, and their relative importance is more a function of the problem being studied than it is of the statistics. Let's suppose that you are a physician, and a new medicine becomes available. You consider whether or not to administer the new medicine to one of your patients. There are two possible mistakes you can make:

1. You can give it when you shouldn't have.
2. You can fail to give it when you should have.

However, either way you must decide between giving the medicine and not giving it. Whatever the statistical characteristics of these two choices may be, the proper decision depends upon an appraisal of the *relative risks* of the two procedures.

If the risk of giving it is an upset stomach, but the risk of not giving it is death, you would likely opt to run the first risk rather than the second. Statistical procedures might be able to quantify the *probability* of one risk versus the other, but your decision would still probably depend on your assessment of the *consequences* to the patient of the two possible choices, as well as what you may know about the statistical probabilities.

Thus, in real life situations there are always two aspects to such decisions: (1) the probability that the risk will occur (a matter amenable to statistical analysis); and (2) the seriousness of the risk if it does occur (amenable only to analysis and understanding of the substantive situation and its consequences), which is a matter to which the statistical analysis is blind.

It is not possible in this book to address problems pertaining to the seriousness of consequences. We must confine ourselves to the consideration of the two types of statistical error. Much of statistical decision-making involves these two kinds of error. From a statistician's viewpoint, a good decision is one that recognizes the potential errors involved and controls them as much as is possible under the circumstances.

Translating all of this into the statistical jargon, remember that the statistical testing procedure is to set up a null hypothesis and test that. Then, the two types of statistical error discussed above *are always phrased in terms of the null hypothesis*. They are:

Type I: Rejecting the null hypothesis when (unknown to us) in reality it is true (False Rejection);

Type II: Accepting the null hypothesis when (unknown to us) in reality it is false (False Acceptance).

Type I is sometimes called "alpha error" or the "level of significance," and is the probability of detecting an effect that really doesn't exist. Type II is sometimes called "beta error," and is the probability of failing to detect a real effect. More about this later.

D. The Role of Probability

I have mentioned that probability plays a critical role in inferential statistics. In essence, the goal of inferential statistics is to make general statements from partial data *with acceptable probabilities of being wrong*.

What sorts of statements? There are several general types of statements that are usually the focus of inferential investigations. These include:

- Does this score come from a distribution in which a certain statistic has a certain value?
- Is this value (of a statistic such as a mean, SD, or correlation coefficient) greater than, less than, or different from another value of the same statistic?
- Is this value different from zero?
- How close is my sample value to the true value of the statistic for the population?

Examples of each of these questions might be, respectively:

- I bought a pair of shoes guaranteed to wear at least 12 months; mine wore out in 6. Is it likely that my shoes were a member of a population of shoes with an average life of 12 months, or is the manufacturer guilty of false advertising?
- I taught one class reading by the phonics method and another by whole word techniques; one scored a mean of 45 on my end-of-year proficiency test, and the other 39. Are these results really different, or is this apparent difference just the result of chance?
- I correlated end-of-semester grades for my 7th graders with body weight, obtaining an r of .20. Does this represent a real relationship, or is it a chance fluctuation where the real relationship in the population at large is zero?
- I surveyed a sample of schools and found that they required on average 1.5 hours of homework per school night. How close am I to the true mean for all schools?

Statisticians contrive to answer such questions by developing a probability statement. You may have noticed that my examples dealt almost exclusively with averages. This is because it is mathematically difficult to develop probability statements for any index that you might wish to compute, and the theory of the field has been developed more completely for statistics of the moment system, the arithmetic mean and the standard deviation, and the product-moment correlation. There are some other tests and variations, including what are called "nonparametric" statistics, but we will not treat these here.

1. The probability distribution

Thinking back to Chapter 3, you will recall that a distribution consists of laying out the possible values of a variable and representing the frequency of occurrence for each. If we then plotted the frequencies for each of the values, we could draw a picture or graph of the distribution. For a single sample or study, this was as far as we could go.

But, if the same study were to be repeated time and time again on different samples, the composite graphs would converge into a smooth curve, approximating the population distribution for this variable. For a few, particular variables, this population distribution has well-known characteristics.

As you may know, any line graph or "curve" can be expressed mathematically as an equation. For a two-dimensional plot, this equation is usually stated as $y = f(x)$, where y stands for a value along the y-axis and $f(x)$ is a shortcut way of saying "some function of the corresponding x (i.e., value on the x-axis)." In other words, for each possible x value, there is a corresponding y value that can be computed by substituting the x value into whatever the function expression is.

When this is done, and the equation of the curve properly processed by mathematical techniques beyond this discussion, certain statements about the probabilities of occurrence of the different values of the variable can then be made. It is a table of such probabilities that is referred to as a "probability distribution." Its usefulness is that it enables you to associate a probability with the values of that same variable when it is obtained from your study.

One problem of course is to contrive to convert the variable that you are interested in into one of those for which a known probability distribution exists. Fortunately, there are several such distributions. Also, mathematical statisticians have devised ways to translate some of the more frequently observed statistical values into variables resulting in one of these distributions. (For example, comparing a difference to the standard deviation of that same difference usually results in a new variable with one of the desired distributions.)

However, in any case, there are several in wide use, for which tabled probability values exist. Among these are:

1. The standard normal distribution—used to test the difference between an obtained mean and a hypothetical mean where the population SD is known (or a reasonable approximation is available). Several other tests approximate this distribution, for large samples: tests between two means, tests of the difference between two correlation coefficients, tests of the difference between two proportions.
2. Student's "t" distribution—used to test the difference between two sample means, and to test whether a correlation coefficient or regression weight is different from zero. ("Student" was a well-known British

statistician, William Sealy Gosset, employed by the Guinness brewery in Dublin. He published a classic paper upon which this distribution is based in Biometrika, 1908.)

3. Fisher's "F" distribution—used to test the difference among more than two means, to test whether the multiple correlation is different from zero, and to test the difference among several variances. (Fisher developed this distribution in connection with his extensive studies of agricultural crop yields.)

4. Chi-square distribution—used to test the difference between expected and obtained frequencies in categorical data.

5. Binomial distribution—used to test the difference between an observed and a hypothetical proportion.

It should be understood that the above list is not exhaustive, particularly with respect to applications.

2. The tabulation of a probability distribution

You may never have occasion to look up a probability in such a table, but it is safer for me to spend a few moments on this subject, just in case.

It is important to first understand that the mathematical process of defining a probability distribution results in a fact that may at first surprise you. That is, that the probability is represented on the graph of the probability distribution not by the ordinate (vertical measurement at a particular point on the baseline), but rather *by an AREA bounded by the curve itself, the baseline, and the ordinates at two points on the baseline* (See Figure 8-1).

It may not be possible to make this point entirely clear, but consider that the probability of a certain score may be defined as the frequency with which that score occurs as compared to the total score frequency of the distribution.

Now, how do we get to this in terms of areas? Think about the case of a frequency histogram. Here the vertical bar erected over a score interval has a height equal to the number of cases found in that score interval. But the height at any particular point on the score scale does not give you all of the frequencies associated with that score interval.

Why? We must remember that the score scale is a continuum, and that cases, or frequencies, are considered to be spread equally across the *entire range* of any score interval, not at any particular point within it. Now it becomes more clear that in order to get the entire contribution of that score interval you must take all of it, not just a single shaft above some part of it. To get all of it, you take the height times the entire baseline for

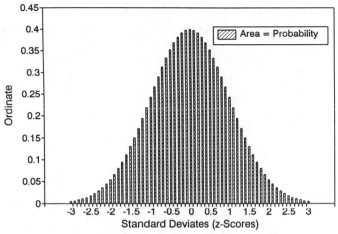

Figure 8-1. Normal Probability Histogram

the interval, or, in essence, its total area. It is the total area of the interval which when compared with the total area in the distribution gives you the proportion of the distribution represented by the interval, or its probability of occurrence.

The mathematics of developing a probability distribution assumes that the area under the mathematically defined curve representing the distribution is equated to 1. This means that, as seems reasonable, the total probability associated with the distribution is exactly 100%.

Because the curve of the distribution is a smooth curve, a series of histographic bars erected along the baseline would not fit it very well. But, if you took those bars more and more narrowly, they would fit the curve more and more closely, and you could add up the areas represented by the bars to get the totality of 100% of the area under the curve. Figure 8-1 shows how a succession of very narrow bars fits the curve quite well. In actuality, the math assumes an infinite number of infinitely narrow bars and adds the area of each together to get 100%.

Because mathematically, in a continuous distribution, a score point has no width (being infinitely small), there can be no areas associated with a single score point. Therefore, any area (probability) must be defined in terms of a range of score points. This is the reason that it is necessary to understand how to use the table of a probability distribution. *It is not possible simply to take a score value to the table and read out how probable that single score value is; one can only read the probabilities of ranges.*

As an example, let's consider the distribution of the standard deviates that we discussed in Chapter 6. You will remember that standard deviates

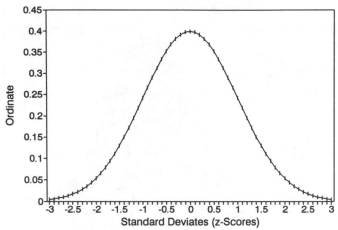

Figure 8-2. Normal Probability Distribution

were the variable formed by subtracting the individual scores in an obtained distribution from the arithmetic mean of the distribution and dividing the result by the SD of the same distribution $[(x - M_x)/\text{SD}]$.

If the distribution of the original variable was normal (bell-shaped), we can refer this standard deviate variable to a table of the probability distribution of the normal curve. (In practice, we do this for any distribution that is a reasonable approximation of the normal distribution.) Anyway, this standard deviate variable is distributed as a standard normal probability distribution with a mean of 0 and a SD of 1.

Figure 8-2 shows the graph of such a distribution with the values of the variable (which is often called "z," or a "normal deviate") marked off on the baseline. As we have previously learned, the normal curve is symmetrical, and centered on its mean. Thus, scores to the left of the mean (0) have negative values, and those to the right have positive values.

Part of a table of the normal probability distribution is reproduced in Table 8-1. We will refer to it several times in the remainder of this chapter.

Now, in reading a table of this distribution, the first trick to remember is that for symmetrical distributions such as this one, the negative side is not usually tabled. Of course the probabilities associated with the negatives are numerically identical to those associated with the corresponding positives, so listing both would be duplicative. *But,* one must remember which sign one is dealing with, as this controls which side of the mean your probability applies to.

Table 8-1. Table of the normal curve

$(X - M_x)/s$	Ordinate	Area*	$(X - M_x)/s$	Ordinate	Area*
0.00	0.3989	0.0000	1.53	0.1238	0.4370
0.03	0.3988	0.0120	1.56	0.1182	0.4406
0.06	0.3982	0.0239	1.59	0.1127	0.4441
0.09	0.3973	0.0359	1.62	0.1074	0.4474
0.12	0.3961	0.0478	1.65	0.1023	0.4505
0.15	0.3945	0.0596	1.68	0.0973	0.4535
0.18	0.3925	0.0714	1.71	0.0925	0.4564
0.21	0.3902	0.0832	1.74	0.0878	0.4591
0.24	0.3876	0.0948	1.77	0.0833	0.4616
0.27	0.3847	0.1064	1.80	0.0789	0.4641
0.30	0.3814	0.1179	1.83	0.0748	0.4664
0.33	0.3778	0.1293	1.86	0.0707	0.4686
0.36	0.3739	0.1406	1.89	0.0669	0.4706
0.39	0.3697	0.1517	1.92	0.0632	0.4726
0.42	0.3653	0.1628	1.95	0.0596	0.4744
0.45	0.3605	0.1736	1.98	0.0562	0.4761
0.48	0.3555	0.1844	2.01	0.0529	0.4778
0.51	0.3503	0.1950	2.04	0.0498	0.4793
0.54	0.3448	0.2054	2.07	0.0468	0.4808
0.57	0.3391	0.2157	2.10	0.0440	0.4821
0.60	0.3332	0.2257	2.13	0.0413	0.4834
0.63	0.3271	0.2357	2.16	0.0387	0.4846
0.66	0.3209	0.2454	2.19	0.0363	0.4857
0.69	0.3144	0.2549	2.22	0.0339	0.4868
0.72	0.3078	0.2642	2.25	0.0317	0.4878
0.75	0.3011	0.2734	2.28	0.0297	0.4887
0.78	0.2943	0.2823	2.31	0.0277	0.4896
0.81	0.2874	0.2910	2.34	0.0258	0.4904
0.84	0.2803	0.2995	2.37	0.0241	0.4911
0.87	0.2732	0.3078	2.40	0.0224	0.4918
0.90	0.2661	0.3159	2.43	0.0208	0.4925
0.93	0.2589	0.3238	2.46	0.0194	0.4931
0.96	0.2516	0.3315	2.49	0.0180	0.4936
0.99	0.2444	0.3389	2.52	0.0167	0.4941
1.02	0.2371	0.3461	2.55	0.0154	0.4946
1.05	0.2299	0.3531	2.58	0.0143	0.4951
1.08	0.2227	0.3599	2.61	0.0132	0.4955
1.11	0.2155	0.3665	2.64	0.0122	0.4959
1.14	0.2083	0.3729	2.67	0.0113	0.4962
1.17	0.2012	0.3790	2.70	0.0104	0.4965
1.20	0.1942	0.3849	2.73	0.0096	0.4968
1.23	0.1872	0.3907	2.76	0.0088	0.4971
1.26	0.1804	0.3962	2.79	0.0081	0.4974
1.29	0.1736	0.4015	2.82	0.0075	0.4976
1.32	0.1669	0.4066	2.85	0.0069	0.4978
1.35	0.1604	0.4115	2.88	0.0063	0.4980
1.38	0.1539	0.4162	2.91	0.0058	0.4982
1.41	0.1476	0.4207	2.94	0.0053	0.4984
1.44	0.1415	0.4251	2.97	0.0048	0.4985
1.47	0.1354	0.4292	3.00	0.0044	0.4987
1.50	0.1295	0.4332			

*Between the mean and this baseline point.

The second trick has to do with the fact that the probabilities, as noted above, apply to variable ranges, not points. In this table (and most such tables) the probability that is tabulated corresponds to the area *between the mean* (0) *and the indicated z value*. Thus, if the tabular entry next to a z value of .75 reads .2734, it means that 27.34% of the total area under the curve lies *between* the mean and the baseline point of (plus) .75! (If it were −.75, 27.34% would lie between the mean and minus .75.)

In terms of probability, the probability of getting a z score between 0 and .75 is .2734, or 27.34%. Another, and perhaps more important, way of reading it is: the probability of getting a z score of .75 *or greater* is the rest of the area above the mean, beyond .75. This is calculated by subtraction: 50% *minus* 27.34 = 22.66%! Correspondingly, the probability of getting a z score of −.75 *or less* is 50 − 27.34 = 22.66%. (The reason that the area beyond .75 cannot be obtained directly, and must be gotten by subtraction, is that this distribution theoretically extends to infinity in either direction. Of course it is not possible to measure from infinity in toward the center point of the curve.)

(Tables of the normal curve are usually given more or less completely. However, for some of the other probability distributions, such as "F," "t," and Chi-square, the distributions are really families of distributions which would take up huge amounts of space to tabulate. In these cases only selected values of the variable that are most frequently used in inferential hypothesis testing are given in the tables.)

III. A Simple Example of Hypothesis Testing

As a simple example of the process of inferential hypothesis testing let's turn our attention to the first of the questions posed above—did my 6-month shoes come from a population with a mean wear-life of 12 months? Let us assume that we know that the SD of wear-life for this kind of shoes from this manufacturer is 4 months, and that wear-life is distributed approximately normally. Then the SD of the mean is $SD_M = SD/\sqrt{N}$, and the variable $x - M/SD_M$ is distributed according to the standard normal curve, where the mean is 0, the SD is 1, and $N = 1$ (just me).

Entering the values for our problem:

$$z = x - M/SD_M = (6 - 12) \div \left(4/\sqrt{1}\right) = -6/4 = -1.5.$$

This finding is to be construed as a value on the horizontal axis (abscissa) for the curve that was shown in Figure 8-2, and the probabilities associated with this finding can be found in the table that was presented as Table 8-1.

Now, as noted above, translating this value to a probability statement is not difficult—but you have to be careful. What is tabulated in the table is not the probability for the specific value -1.5, because mathematically this is unobtainable. What is shown is the probability of getting a value between 0 and 1.5, or, to put it another way, a value as extreme as 1.5, *given that the true value is zero* (that is, that 12 months is actually the true mean). Note that the minus sign simply tells you whether you are above or below the mean (our 6 months of wear was below the advertised mean of 12, hence the minus sign).

The tabulated value opposite 1.5 is .433, which means that if the true mean is 12 months, there is a 43.3% chance of getting a wear value as far below the mean as 6 months, or, conversely, only a 6.7% chance of getting a wear value this low or lower *by chance, where the true mean is 12 months*.

The key phrase is "by chance." The probability distribution really assumes that for a given average, here 12 months, the individual values in the distribution will be spread out according to the standard deviation. (The real reasons that the individual values in any distribution vary among each other are unknown, but *are assumed to be unrelated to the measurement under study—hence, they are collectively referred to as "chance factors."*)

Thus, some 6.7% of the time you would expect to get wear times as low as 6 months or lower from a distribution where the manufacturer's claim of a 12-month average wear-life was really true. The key question now is whether you decide that the shoes came from such a distribution, or, alternately, that they really came from a distribution with a lower mean (wherein your 6 months wear time would be somewhat more likely).

This is a judgment call. The null hypothesis is that there is no difference between the obtained distribution and the advertised distribution. If you decide there is a difference, you know that your probability of error (false rejection) is 6.7% (because 6.7% of the shoes from a distribution with a mean of 12 months would have wear times of 6 months or less).

Statisticians usually adopt a standard in advance of the study in which they argue that they would be willing to accept a certain risk of falsely rejecting the null hypothesis. The most usual values are 5% and 1%. Because the rejection of the null is interpreted to signal support for some hypothesis of interest, which is usually the focus of the investigation, the rejection of the null is called a "significant" finding, and the probability of wrongful rejection (Type I error) is called "the level of significance." (So, at long last, we now know what statisticians mean when they speak of "significance!")

If you had adopted one of the conventional levels of significance in the problem of the shoes, say 5%, you would have to accept the conclusion that your shoes, although of relatively short life, were not so extreme as to justify a claim that the manufacturer was lying about his product in general. (Statistically, you would accept the Null.)

(It is worth noting that the probability distribution used for testing is always centered or based upon the null hypothesis. This is because according to statistical logic it is always the null which is tested. In this way it becomes possible to examine the obtained results in the context of conditions as they would be if the null hypothesis were true. Then the obtained results can be evaluated as being either consistent with or not consistent with the null.)

Now we can state the essence of inferential hypothesis testing. It is simply answering this question: *"Is the observed result so extreme, as compared to the null hypothesis, that we are not willing to accept the probability that it occurred in a distribution where the null hypothesis was really true?"*

If the probability that the observed result came from a distribution based upon the null hypothesis is too small to be acceptable, we reject the null hypothesis, with that same probability (level of significance) of having done so wrongly.

IV. More About the Null Hypothesis

Suppose, as in the case of the shoes, we accept the null hypothesis. What does this mean? Well, (I guess I'm harping on this) it *does not* mean that we have proved the null! (Remember, statistical inference does not permit the proof of *any* hypothesis.) We have merely failed to discredit it. There is a difference. In essence, failing to discredit the null simply leaves the matter open. Perhaps with more cases, fresh data, a more powerful test, etc., we would be in a position to reject it. It is this possibility that keeps graduate students going!

Let me also focus on the problem of why we can't prove hypotheses once more. Part of the difficulty about proving a hypothesis lies in the fact that proving a hypothesis requires demonstrating that no alternative hypothesis can be true. In the case of the shoes, for example, proving the null requires you to be able to say that the mean wear time is exactly 12 months, not 11.99 or 12.01, or any other value. To make such fine discriminations would require prohibitively large numbers of observations, far beyond any practical procedure. So we see again the logical soundness of focusing on discrediting a contradictory rather than attempting direct proof of any hypothesis of interest.

A. Accepting the Null, Power, and Type II Error

Whenever you decide to accept the null, that there is no difference, you run a risk of (Type II) error, but we generally do not have the information necessary to estimate the size of this error. Type II error is the probability of accepting the null when some other hypothesis is true. It is defined as 100% minus the "power" of the test being used.

(As you are not going to find much reference in the reports that you read and the data/information you consume to the concept of statistical power, I am going to put a very brief discussion of this topic in the adjacent box, and move ahead.)

Power is defined as the probability that the test will reject a hypothesis which is untrue. For example, a "powerful" test would be one which would reject the null of 12 months in our shoe example 95% of the time, if the real value were quite close to it, say 11.5. In such a case then, we would estimate Type II error as being 5% (100% − 95%).

A less powerful test might reject the null only 90%, or perhaps 85% of the time under these conditions. Or, it would have a relatively low probability of rejecting the null of 12 if 11.5 were true, but would have an acceptable probability of rejecting the null if the true value were, say 11, or 10.5. And so on.

Obviously, under a given set of conditions, the power of the test increases as the difference between the tested hypothesis and the true value increases. There is a much higher probability of rejecting a Null hypothesis when the true value is a widely different value than there is when it is quite close.

When the true value is very close to the null, the power of the test approaches the level of the Type I error. When the Null is actually the true value, there is no Type II error (you cannot falsely accept the null when it is in fact true). At this point the concept of power loses meaning; and the overall error is simply the level of significance (the probability of falsely rejecting the null when it is in fact true).

Power is more than just setting some arbitrary error level. It is a function of the particular statistical test being used, the number of observations in the computation, and the level of significance (Type I error) chosen. In general, the larger the number of observations, particularly in small populations, the greater the power; and the more Type I error that is acceptable, the greater the power. Of course, as is clear from the above paragraph, the greater the power, the smaller the Type II error.

Although, as indicated above, statisticians rarely calculate Type II error in a particular study, it is important to maintain an awareness of its presence. In some studies it might well have more serious consequences than Type I error. The medical example I gave earlier would be such a case; failure to give the new medicine would be a serious error if the medicine were lifesaving, and the side effects were only headaches.

In addition, there is frequently great temptation to choose more and more stringent controls on the "level of significance" in an effort to be "professionally conservative." After all, what is to be lost if one is conservative and chooses the 1% level of significance instead of the 5% level? Well, as has often been observed, "There ain't no noon repast without remuneration!"

As may have occurred to you from the above discussion, the more stringently you control Type I (choosing smaller and smaller probabilities of false rejection), the more you open yourself to the possibility of Type II (false acceptance of the null). There is no pat answer as to just how to treat this problem of choosing the appropriate size of the significance level, but it is advisable to refrain from the simple course of making Type I error, the level of significance, smaller and smaller without limit.

To compensate for this shortcoming, researchers typically attempt to formulate their problems for test in such a way that the relative importance of controlling Type I error outweighs that of controlling Type II error. However, it is easy to get into the habit of forgetting all about Type II error. The good statistician and researcher must constantly remind himself that setting the level of significance (Type I error) is not all there is to controlling the error in his study. Also, it is not necessarily conservative to move from an alpha level of 5% to one of 1%!

B. One-Tailed and Two-Tailed Tests

There is one more piece of statistical logic to be dealt with here. That is the matter of directionality in hypothesis testing. In the example of the shoes, above, it was clear that my concern was whether or not the manufacturer was overstating the average wear time of his product. I would not have been concerned if my shoes had worn *longer* than his published mean wear time; I only wanted to be sure that I wasn't being misled or cheated by believing his wear time claim.

In such a situation, where we are only concerned with significant departures from the null *in one direction* (either up or down), we say that we are going to test the proposition with a "one-tailed test." In doing so, we take the position that we don't care about differences in the other direction, and concentrate the test on the direction of concern. Therefore, such tests cannot pick up unexpected differences in the other direction.

The one-tailed test is accomplished by focusing the significance level in one direction. Remember that a significance level of, say, 5% means that we will accept the probability of falsely rejecting the null five times out of 100, when it is actually true. By doing a one-tailed test, focused on the possibility that the shoes actually came from a population with *less* than 12 months mean wear, and ignoring any possibility that they came from a population with a mean higher than 12, we can concentrate all of our level of significance probability at the critical end of the value scale where we are most concerned.

Consider, we have noted above that in a probability distribution, such as those we use to test hypotheses, probability is represented by the area under the curve between two points on the baseline. What I want to do is to pick a point on the baseline so that anything more extreme than that leads to a conclusion that the null hypothesis is unsupportable.

I define unsupportable in terms of the fact that the probability that a value so extreme could occur when the null is true is too small to believe. Now, if I put all of my probability on one side of the distribution (see Figure 8-3 which shows a 5% one-tailed test on the right and a 1% one-tailed test on the left), and ignore the other, I have a more sensitive test of the null hypothesis than I would otherwise have. That is, I will be able to reject on the basis of a smaller departure from the null than otherwise.

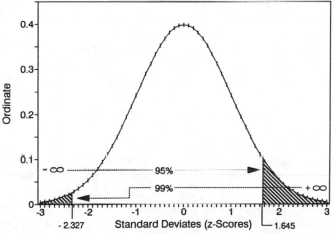

Figure 8-3. One-Tailed Tests (Normal Distribution)

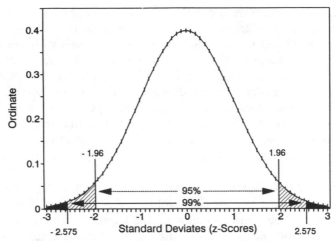

Figure 8-4. Two-Tailed Tests (Normal Distribution)

In contrast, if my problem is to determine if the observed result is *different* (*either* bigger or smaller), I can't concentrate my test on one side or the other, *but must split my probability of false rejection between the two sides.* If I am only willing to accept the probability of falsely rejecting the null of 5%, but I have to be able to detect observed results that are bigger as well as those that are smaller, I am going to have to divide my possible errors (significance level) *half to each side.*

This will result in a less sensitive test on either side (the difference of the observed value from the null will have to be bigger to be detectable). However, as compensation, I won't miss a result that is in the direction opposite from what I might have anticipated. (See Figure 8-4).

Keeping the general layout of Figures 8-3 and 8-4 in mind, let's take a numerical example, again using values from the table of the normal curve (Table 8-1). Suppose I want to do a one-tailed test at the 5% level, focusing on values above the null. Then my entire probability is located above the null, and I can detect a value of 1.645 or larger on the baseline, i.e., $(x - M_x)/SD_M \geq 1.645$. This is because the table says that 45% of the normal distribution lies between its mean and $+1.645$, so that 5% must be associated with values more extreme than $+1.645$!

However, I cannot tell anything about values that are below the null. I have used up all of my test on the upper side. Figure 8-3 shows the critical areas for a 5% upper and a 1% lower one-tailed test.

On the other hand, suppose I decide that differences in either direction from the null might be important. To do a two-tailed test, I must split my 5% probability of error to cover values above and below the null, 2.5% to each. Again referring to the table, I see that 2.5% (read in the table as

.475, since the table gives the proportion of the curve between the mean and the baseline point) is associated with a baseline value of 1.96. So, I must have a value of 1.96 or more extreme to reject the null. But, *it is 1.96 or bigger on the one side, or −1.96 or smaller on the other*, because of the two directional nature of the problem! Figure 8-4 shows the critical areas for 5% and 1% two-tailed tests.

In a sense one-tail and two-tailed tests are equally sensitive (e.g., if they are at the 5% level, each will reject the null hypothesis falsely 5 times in 100). However, the former is totally insensitive to differences in the unexpected direction, whereas the latter is equally sensitive to differences in either direction, though only half as sensitive to either as the former is to the direction it is designed for.

C. Some Examples and Cautions

Let's take some examples of situations that might logically call for one-tailed tests. The case of the shoes is one. Another might be, testing for the average breaking strength of bridge steel—we don't care if it's too strong, but it is important to know if it's not strong enough. Another might be testing for food contaminants—here we don't care if the contamination level is too low, but it's critical that it not be over the production standard. Etc.

Now let's look at logical two-tailed situations. Suppose you are the producer of 1/2-inch bolts. In your quality control tests, you will use a two-tailed test, because it's just as bad to produce 1/2 inch bolts that are excessively long as excessively short—either is a quality control problem. Or, suppose you are trying a new teaching technique. Of course, you hope it will result in improved learning, but you can't afford to ignore the possibility that it may inhibit students' progress. You had better use a two-tailed test.

Or, as financial manager to a mutual fund you tend to favor "blue chip" stocks. So, you structure two portfolios, one of "blue chip" stocks, and one of "second tier" stocks as a test of your procedures for maximizing returns. The following year, you should check their relative performance with a two-tailed test of stock performance—it could be expensive to test only how much better blue chips have done if the opposite were to be true!

Like the choice of the significance level, choice of one vs. two-tailed test should be made prior to collecting the data. Both should grow naturally out of the logical consideration of the problem. This is because the probabilities listed in the tables of probability distributions are constructed under the assumption of no knowledge of the results of the investigation prior to conducting the statistical tests.

In addition, it is wise to remember that all data collection involves errors. Data cannot be used repeatedly to search for significances. The repeated use of the same data tends to capitalize on whatever errors were present, and there are always errors present! The net effect of this tendency on the true probabilities is unknown, but probably important. Thus, findings that are teased from repeated manipulations of data must be verified on fresh data to be acceptable.

Indeed, it is a canon of the scientific method that results be independently replicated and verified in any case. This is the best and ultimate definition of "significance."

And, also *remember that even if fresh data are used, you will get a falsely significant result 5 times out of 100 anyway*, given the 5% level of significance, since in a large number of efforts that's how often you will reject the null, even when nothing is going on!

D. The Concept of Degrees of Freedom

There is one more wrinkle in testing hypotheses. You will recall that I noted above that some of the standard probability distributions were really families of distributions; and that in such cases only a couple of points from each possible distribution would be found in their tables.

I did not specify at the time what distinguished the members of these families from each other. Now I can no longer avoid it. The distinguishing characteristic is called "degrees of freedom." Degrees of freedom is a notion that results from constraints imposed by the mathematics of deriving the probability distributions associated with the variables which we contrive for hypothesis testing.

I shall not attempt to take you through the niceties of deriving these many distributions, but we should take a look at the concept.

The key idea in the concept of degrees of freedom is independence of information. Whenever knowing one fact about a situation predetermines another, there is really only one independent fact, and thus "one degree of freedom" in the situation. Take the following example comprised of three "facts":

1. My income last month was $1000.
2. My expenses last month were $800.
3. My bank balance at the end of last month was $2500.

In this case, knowing any one or two of these three pieces of information cannot lead you to any of the others. Thus, these three pieces of data are in fact independent, and we would say that we had 3 degrees of freedom.

Now, let's substitute the following "fact" for the third one above:

3a. My profit last month was $200.

Now, in this new case, it is clear that if we know any two of these pieces of information, the third item is fully determined. Knowing that profit was $200, and income was $1000, we can derive expenses without any further information. Or, knowing profit and expenses, we can calculate income of $1000, without being given the third fact. Thus, instead of having three independent pieces of information (3 degrees of freedom), we only have two—any of the three statements is fully determined by the other two.

For a "legendary" example of degrees of freedom see the adjacent box.

A father left his estate to whichever of his two sons was the better predictor. A test was set up comprised of tossing a coin. The first son, Algernon, tossed the coin three times, correctly calling: "Heads will be up"; "Heads will be down"; "Tails will be up". Smiling, he handed the coin to his brother, Plimpton. Plimpton took the coin and tossed it, saying, "Heads will be down; Tails will be up". This was true. Again he tossed it, saying, "Tails will be up; Heads will be down". Again, true.

At this point Plimpton handed the coin to Throttlewattle, the estate lawyer, saying, "I have called the coin correctly four times to my brother's three. I gratefully accept my father's estate." However, Throttlewattle (who had read this book) said, "Not so fast! Although you made four calls of the coin, you made two calls on each toss. Each time, one of those two calls was completely dependent on the other. Therefore, your predictions involved only two degrees of freedom, or two independent calls. Algernon, on the other hand made three completely independent calls (3 degrees of freedom), and thus is the better predictor, 3 to 2."

Let's do it again, just with numbers. You can draw 10 numbers, but you are limited by the fact that they must add up to 25. You can literally draw any numbers, or choose any numbers, you wish for 9 trials. But the 10th draw is not free. It is whatever number which, when added to the sum of the first 9, makes 25. Thus, you have only 9 free choices in these 10 numbers. Statisticians would say that you have $N - 1$ degrees of freedom here ($10 - 1 = 9$).

Let's take a quick look at a case of two constraints. You may select any 6 positive numbers, call them $X_1 \ldots X_6$, but it is required that $X_1 + X_2 = 13$, *and* that the grand total equal 50. Thus, the first number fully determines the second, and the second number has no freedom. The selection of numbers 3, 4, and 5 is again free, but X_6 is now fully determined, since it must be whatever will make the grand total 50. So, we have a degree of freedom for the first, third, fourth, and fifth numbers in this set, or 4 degrees of freedom. This is equivalent to the original 6, minus the two constraints $(N - 2)$.

Degrees of freedom is an important concept. It should be intuitively clear that the greater the number of independent facts that are known about a situation, the more the useful information. On the other hand, it is seriously misleading to believe that one has a large fund of information about a subject if most of those facts are really dependent on each other (the same bit of information in different guises).

From the standpoint of statistical procedure, the importance of this concept arises from the fact that when we contrive to convert our statistics, or variables-of-interest, into variables with known probability distributions, the process sometimes results in variables that are sensitive to the constraints inherent in the statistics with which we work. Then the tables of their probability distributions must take these constraints into account.

In such tables, it turns out that there is a separate distribution for each degree of freedom, or combination of degrees of freedom, creating the families of distributions that I referred to earlier. In essence, standard distributions such as chi-square, t, and F differ as a function of the number of degrees of freedom involved in the test.

As an example, consider the variance of a distribution. You will remember that the variance is based upon (the squares of) the deviations of the scores from the mean of the distribution. Those deviations have a constraint built into them as a function of the way the mean is defined; it is that the sum of the deviations from the mean must be zero.

Thus, no matter how many deviations there are in your study, there is freedom of choice about only $N - 1$ of them; the last one is always determined because it must be whatever is necessary to cause the sum of the other $N - 1$ of them to add to zero! *As a result, the variance and the standard deviation of a sample distribution have $N - 1$ degrees of freedom!*

As a second example, let's consider the chi-square table that we looked at in the last chapter (see Table 8-2):

In a contingency table such as this one, we have one element of the chi-square computation for each of the cells, comprised of a value resulting from $(f - F)^2/F$. The marginal totals are taken as a given. Now, the question is how many of these cell computations are based on free choice? Well, it doesn't make any difference whether you talk about the "expected" frequencies (F) or the actual frequencies (f); in either case they must add to the marginal totals.

Therefore, each row has only 1 degree of freedom, since the other cell is determined by the necessity of adding up to the marginal value. Thus, the entire second column is eliminated. Looking at the first column, there are only two free choices, since the third is determined by the necessity of

Table 8-2. Contingency table for chi-square analysis

SES	Voted		
	Yes	No	Total
High	8	7	15
	(8.4)	(6.6)	
Middle	15	5	20
	(11.2)	(8.8)	
Low	5	10	15
	(8.4)	(6.6)	
Total	28	22	50

adding to the column total. As a result, *you have only 2 degrees of freedom in this entire table*. The other 4 cells are determined by the need to add to the respective marginal totals.

In general, in tables of this sort, you lose a row and a column. The number of cells left, that do have freedom, *can be summed up as the product of the number of rows less one times the number of columns less one*. Here this would be $(3 - 1)(2 - 1) = 2$.

To evaluate the significance of the computed chi-square value here, you would have to consult the probability distribution table specifically for values of chi-square distributed with 2 degrees of freedom.

In most but not all applications, chi-square has either $N - 1$ or $(r - 1)(c - 1)$ degrees of freedom, where N is the number of categories within which the frequencies are distributed. Student's t usually has $N - 1$ degrees of freedom (where N is the number of observations), but, depending on the situation may have $N_1 + N_2 - 2$, or some other number. Fisher's F distribution arises from the ratio of two independent estimates of the variance of a population, and has $N - 1$ degrees of freedom associated with its numerator estimate, *and* $N - 1$ associated with its denominator estimate.

(Note that these two Ns may be different since they are the Ns associated, respectively, with the numerator estimate of variance and the denominator estimate of variance.) Thus, the F probability tables differ for each *combination* of numerator and denominator degrees of freedom. (Now, you can imagine why it would be extremely cumbersome to try to publish entire distributions for this test!)

1. Form of the tables

There is a different probability distribution for each degree of freedom or combination thereof, but to save space, only key values of the distribution are given in many tables. It works like this:

For chi-square and t, one dimension of the table will be number of degrees of freedom, and the other will be selected probability points, such as the 90% point, 95% point, 97.5% point, 99% point, and 99.5% point. In the body of the table will be found the baseline values of the statistic corresponding to each of these points. Thus, the interpretation at the 95% point is that the value in the cell is a value that could be equalled or exceeded only 5% of the time if the null hypothesis were true. This would correspond to a 5% one-tailed test or a 10% two-tailed test, given the indicated number of degrees of freedom.

As another example, if you wanted a 1% two-tailed test, you would have to have a value such that half of that area would be on either side of the curve. Thus, for the correct number of degrees of freedom, you would look up the tabled value at the 99.5% point. If your computed result equalled or exceeded the tabled value at that point, you would reject the null hypothesis with significance level of 1%, two-tailed, and that same probability of false rejection.

For the F test, you would find the numerator degrees of freedom listed across the top of the table, and the denominator degrees of freedom down the side. In the body of the table you would find two values for each cell, that associated with the 95% point and that for the 99% point. Thus, probabilities would be available at only these two points on the curve. Again, if the computed variable equalled or exceeded the tabled value the null is rejected, with a probability as indicated if it was, as usual, a one-tailed test; but, depending on the logic of the problem, twice that if a two-tailed test.

V. The Analysis of Variance

The reader must refer to one of the listed references to get a broad picture of this topic, but we cannot avoid the need to lay out at least the basic concept of analysis of variance (ANOVA). I touched very briefly on this area in the preceding chapter in connection with the analysis of covariance, with which it is usually linked.

Analysis of variance is a series of techniques and situations that has very broad application in the arena of inferential statistics. It is focussed largely on the problem of detecting mean differences through the analysis of the total variation in a data set according to the sources which gave rise to that variation. Basically, the concept postulates that this total variation in the data can be partitioned in such a way that portions of it can be attributed by inference to specific sources, with specified probability.

Of course, the classic research situation is to compare the performance of an experimental group to a control group in order to see if some experimental treatment made a discernible difference in performance. Frequently we use the arithmetic mean on some performance measure as the statistic to be compared. The standard inferential comparison would employ the "t" distribution to examine the null hypothesis that these two groups did not differ in mean performance.

But suppose you wish to test the difference among more than two means? For example, among three different treatments and a control? Unfortunately, we cannot use "t" in this situation, as it is no longer possible to structure this situation as a variable with a probability distribution corresponding to "t." It is this sort of situation that requires the ANOVA approach. The additional reach and flexibility of the ANOVA procedures make it possible to subject a wide variety of situations with much greater richness and portent to inferential analysis.

You might object that if "t" is limited to testing the difference between two means, the same purpose could be served by applying the "t" test to successive pairings of the four sets of data (3 treatment means and a control mean). Yes, this can be done, and the resulting 6 "t" tests examined.

But, there are several objections to doing it this way. First, these tests are not independent of each other, since each set of data is used several times, once each in combination with the various other sets. This proscribes the use of "t," since the derivation of its distribution assumes that each mean tested is obtained independently of the others in the analysis.

Another difficulty is that when you make a large number of such "t" tests, you can expect a considerable number of them to come up significant when absolutely nothing is going on (even more than the probability of Type I error, because of the interdependency). Finally, in some applications of ANOVA it is important to examine something known as "interaction," which cannot be done through pairwise testing. So, it is very valuable to have a way to examine more than two means simultaneously.

A. A Simple Example of ANOVA

Although it is not my purpose to stress computation in this text, I believe that we will have to go through a simple ANOVA problem to get the idea across. Suppose we take the very simplest application possible, comparing the means of three groups, A, B, and C on a single variable. For simplicity, let's say that there are four persons in each group. The data are given in Table 8-3. (Note that d^2 refers to the squared deviation of the score from the mean of its own group, while d_g^2 refers to the squared deviation of the score from the grand mean of all scores.)

The single most important fundamental fact in ANOVA is that the sum of the squared deviations of the individual observations from their grand

Table 8-3. Table for simple analysis of variance

Group A			Group B			Group C		
X_1	d^2	$d_g^{\,2}$	X_2	d^2	$d_g^{\,2}$	X_3	d^2	$d_g^{\,2}$
1	4	10.9	2	4	5.3	1	25	10.9
3	0	1.7	4	0	.1	5	1	.5
3	0	1.7	4	0	.1	8	4	13.7
5	4	.5	6	4	2.9	10	16	32.5
12	8	14.8	16	8	8.4	24	46	57.6
M	3		4			6		
V		2			2			11.5

Grand Mean (M_t): $(12 + 16 + 24)/12 = 4.3$
Total Variance (V_t): $(14.8 + 8.4 + 57.6)/11 = 7.35$

mean can be partitioned into additive subsets that can be associated with specific sources of variation. Furthermore, our mathematical statistician friends have shown that these component subsets can be made into *independent* estimates of the population variance, in spite of the fact that they are derived from the same data set. Thus, the ratio of such independent estimates may be taken and the result referred to the probability tables for Fisher's F distribution.

Now, it is not actually the variance itself that is analyzed, but rather the numerators of the variances. This is acceptable, because the essential information is contained in the numerators. For data drawn from samples, these numerators, or sums of squared deviations, become estimates of the variance when they are divided by the appropriate number of degrees of freedom. These are typically the number of elements that went into each respective sum of squares, minus 1. (In our early discussions we have used N as the denominator for the variance, not $N - 1$. The difference arises from the mathematical constraints I referred to above, and should not trouble you overly much.)

Let's see how this works with the data in Table 8-3. The total sum of squares (the numerator of the total variance) would be $14.8 + 8.4 + 57.6 = 80.8$.

However, we might also estimate the population variance from the variability shown within the three groups vis-a-vis their own means. Adding up the sums of squares for each group gives: $8 + 8 + 46 = 62$.

Finally, considering our three group means, under the null hypothesis that they do not differ from each other, and thus must each be an estimate of the population mean, their variation among each other gives us another estimate of the population variance. If we assume that the first group is composed of 4 scores of 3 (its mean), the second is composed of 4 scores of 4 (its mean), and the third is composed of 4 scores of 6 (its mean), then we can get the sum of squares of the deviations of each of these scores from the grand mean (4.3): $4(3 - 4.3)^2 + 4(4 - 4.3)^2 + 4(6 - 4.3)^2 = 18.7$.

Looking back, we see that we had a total sum of squares of 80.8, which we now find is comprised of a "within groups" sum of squares of 62 and a "between means" sum of squares of 18.7. (The partitioning is off by .1, which is simply accumulated rounding error.)

Since our F test requires the ratio of variance estimates, we must now convert our sums of squares into estimates of the variance of the population represented by the 12 scores. As mentioned earlier, the appropriate procedure is to divide each sum of squares by its corresponding number of degrees of freedom. (This is important, theoretically, but particularly in the case of small samples where there is considerable numerical difference in dividing by N, as opposed to $N - 1$.)

The only constraint on our "total variance" is that the deviations must add to zero (as we noted earlier), so the total sums of squares should be divided by $N - 1$, or 11, to get the total variance: $80.8/11 = 7.35$.

The "between means" sum of squares involves only three different values, and these three are constrained by the fact that the sum of their deviations around the grand mean must equal zero. Thus, the appropriate number of degrees of freedom here is the number of groups, minus 1, or 2. The variance estimate becomes $18.7/2$, or 9.35.

The "within groups" sum of squares involved N observations, but there was one constraint within *each* of the groups, that is, the sum of the deviations around the group mean be zero. Thus, the appropriate number of degrees of freedom for this variance estimate is N, less the number of groups, or $12 - 3 = 9$. Then the "within groups" variance estimate becomes $62/9 = 6.89$. (*Note that the number of degrees of freedom for the partitioned sums of squares MUST add to the total number of degrees of freedom for the entire data set $(2 + 9 = 11)$.*)

All that remains to test the null hypothesis of no difference among the three group means is to take the ratio of our two independent variance estimates, the "between" and the "within." The former is the treatment condition (that is, if the treatment had made enough difference to be detectable, we would expect the means to differ so much that the variance as estimated from them would be greater than that estimated from within the groups.)

To avoid values of less than 1, we always place the variance estimate expected to be larger (that is, the "between" variance) in the numerator of the F fraction. Thus, in this example, the F-test variable is $9.35/6.89 = 1.36$.

To ascertain whether or not the computed F is so large as to make it improbable that both estimates were estimates of the same population variance, we turn to the F tables with numerator degrees of freedom of 2, and denominator degrees of freedom of 9. The tabled values for this cell are 4.26 and 8.02, representing the 5% level of significance and the 1% level of significance, respectively. Since our computed value of F did not exceed either one of the two tabled values, we must accept the null hypothesis that our three group means do not really differ among themselves.

Because the tables don't give the entire F distribution for degrees of freedom of 2, 9, we cannot tell where the value of 1.36 is in the probability distribution. However, because it is not close to the value at the 95th percentile of the distribution, it is clear that we have not come very close to rejecting our null hypothesis with an acceptable probability of being wrong.

Of course, remember, we have not proved the null either, and we had a very small number of observations. With a larger study and many more observations we might have been able to detect differences that are not apparent at this level of inquiry. (The reader may have noticed before that the more observations that are taken in a research study, the better the chances of identifying subtle and small effects. This tends to be true in most statistical studies.)

B. More Complex Examples of ANOVA

We are not going to go through more complex ANOVAs in detail. But I want to give you some idea of the richness of application of this basic approach.

First, let me say that, fundamentally, however complex the details, all of the ANOVAs work more or less like the example we went through above. The total sum of squares for the data set is partitioned into several additive components, each of which when divided by the appropriate number of degrees of freedom becomes an independent estimate of the variance. Then estimates associated with the treatment effects of interest are compared to other estimates by using the F distribution, with the indicated number of degrees of freedom. If the computed F value exceeds the tabled value, then it is concluded that the treatment effects are too large to be consistent with the null hypothesis.

Our example above is sometimes called an example of "one-way ANOVA." This is because there is only one variable involved, and the means of some number of groups on this single variable are the subject of the analysis. There are a number of variations in which there are two (or more) variables involved. One such variation might be as follows:

Suppose there are three groups of reading instruction. But we are interested not just in the relative mean performance of students as a function of the way they were taught. We are also interested in the socioeconomic background that they bring to the instruction, as a given instructional method might work differently for students of differing background levels.

Thus, groups of students might be classified simultaneously with respect to both variables, socioeconomic level, and reading instruction method. If we had, say, three types of reading instruction and four levels of socioeconomic background, we would have a 3 × 4, two-dimensional classification with a total of 12 cells. If we take the simplest version of this design, we would assign equal numbers of each background group to the three reading treatments randomly, conduct the instruction, and then examine the reading performance means.

However, the fact that we have two variables involved in this design results in two null hypotheses: First, that the row means (socioeconomic categories) do not differ; and second, that the column means (reading treatments) do not differ.

C. Interaction

In fact, there is actually a third null hypothesis involved: that the "interaction" is zero! We need to digress here for a moment to discuss the concept of interaction.

In the experimental design, we are usually mostly interested in what are called the "main effects." In this case that would be primarily identifying mean reading performance differences associated with instructional method, although our design also enables us to ask if there are mean reading differences associated with socioeconomic status. Either of these would be considered "main effects."

(Also, the layout of the design could make either one the row variable and the other the column variable. In addition, there is no theoretical restriction on the number of rows or columns possible; though, practically, in order to have clear cut effects, it is well to limit the number of cells in each dimension to a few well-defined categories. Practical sample sizes may also constitute a limitation. It is well to have no empty cells, and things are generally easier if there are the same number of observations in each cell.)

The concept of interaction recognizes the fact that the relationship between two variables, such as reading performance and instructional method, may be different as a function of another variable, such as socioeconomic status.

For example, if we found that the means for Method 1 and Method 2 were significantly different, in the same direction and by comparable amounts, no matter which level of socioeconomic status we were looking at, then we would conclude that no interaction existed. However, if we found that Methods 1 and 2 differed significantly for high socioeconomic status students, but not for low status students, we would have to conclude that the treatment effect was dependent on socioeconomic status—that there was a "significant interaction."

Let me hasten to emphasize that the main effects may still be significant whether or not there is a significant interaction, but the interpretation of the findings is somewhat different. The finding of a significant interaction sharpens the results by suggesting delimitations of the main effects which should be taken into account in applying the results of the research.

Significant interactions can be identified statistically in the ANOVA process, of course. But, particularly in the case of simple interactions, plotting the subgroup means reveals the differences nicely. Let's continue our example, not numerically, but in schematic form. Suppose we take Low, Medium, and High on the SES dimension, and Experimental (receives a treatment of some sort) and Control (no treatment) on the other dimension. We might then plot the subgroup mean scores in order to examine the interaction, if any, of treatment effects with SES status.

Figures 8-5 through 8-8 illustrate the way plots of simple interactions might look under conditions of several possible outcomes of such a study.

Figures 8-5, 8-6, and 8-7 are examples of significant treatment effects. In the first of these there is an interaction, because the treatment apparently works for lower SES children, but not for upper level children. In the second, the interaction is more dramatic, because the success of the treatment is actually inversely related to SES status. In the third case,

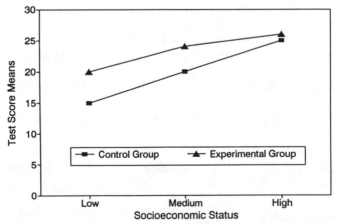

Figure 8-5. Interaction Plot: Subgroup Means

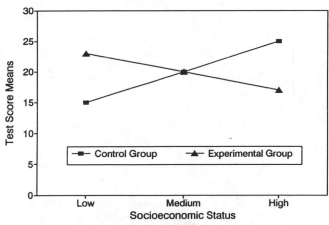

Figure 8-6. Interaction Plot: Subgroup Means

the treatment effect is equally significant at all levels, and no interaction is found.

In the last figure, Figure 8-8, there is no treatment effect to be found, and thus no treatment by SES interaction, although it should be noted that there is a clear, curvilinear relationship between mean score and SES status for both Experimental and Control groups.

In a study such as the one for which we have just looked at possible interaction plots, the tests of the null hypotheses are accomplished as usual. The total sum of squares in the data set is partitioned into additive

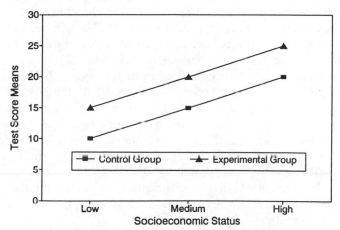

Figure 8-7. Interaction Plot: Subgroup Means

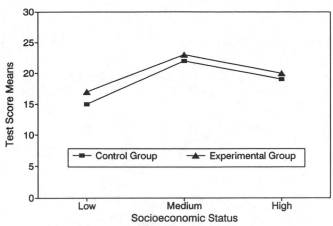

Figure 8-8. Interaction Plot: Subgroup Means

components associated with the rows, with the columns, with the interaction, and with the remainder (called the "within" or "error" sum of squares). Again, the total number of degrees of freedom $(N - 1)$ is also partitioned similarly.

If we had had 10 cases in each of our 6 cells, or groups, the total number of degrees of freedom would have been 6×10, less 1, or 59. The number of degrees of freedom associated with the "within groups" or "error" sum of squares is the number of cases less one in each group summed over all groups. In our example, then, there would have been 6 times $(10 - 1)$ or 54 degrees of freedom there.

The main effects degrees of freedom are simply the number of categories less one in each case, which arises from the constraint associated with the row mean or column mean, respectively. Thus, in our example, we have $3 - 1$, or 2, for the socioeconomic status degrees of freedom, and $2 - 1$, or 1, for the Treatment degrees of freedom.

In this example the interaction degrees of freedom is 2. This is derived from the fact that the interaction considers all of the row cell means against their row means and all of the column cell means against their column means. However, when you have examined 2 of the row cells against the row mean, the third is determined, so you lose a degree of freedom; similarly, you lose a degree of freedom in considering the column cell means. Thus, similarly to the chi-square contingency table, you have a set of "free" cells equivalent to the number of rows less one times the number of columns less one. In our case this is $(3 - 1)(2 - 1) = 2$.

At this point you should note that the degrees of freedom for the various components add up to the total degrees of freedom in the problem, just as they are supposed to. The total is 59; and we have 54 for error, 1 for columns, 2 for rows, and 2 for interaction.

Independent estimates of the population variance are formed by dividing the component sums of squares by their respective degrees of freedom. The ratios of the effects variance estimates are taken with the estimate derived from the "error," and the results referred to the F tables with the appropriate numbers of degrees of freedom.

D. Some ANOVA Variations

It is possible to vary the two-way (sometimes called "factorial") design in several ways. For example, in the above case, we might have exposed each student to several instructional methods. However, this creates differences in the theoretical model of the design. First, we no longer have independent samples of students, which means that any peculiarities of the students apply equally to all instructional methods. This means that the results of comparing the methods cannot be freed of the influence of student characteristics, as it formerly was by the random assignment of students to treatment conditions.

In addition, such a change also creates problems of order. To what extent are the results influenced by the order in which each student receives the reading instruction methods? Fortunately, variations of ANOVA exist that can deal with these matters.

Other variations include various ways to assign subjects to treatments combined with ways to replicate treatments or subjects in various orders and subsets. We need not consider these in detail here. The reader is referred to the reference list.

In general, ANOVA deals with two or more sources of variation arising from: one or more treatments, individuals, orders of treatment, replications, interactions among these various factors and otherwise unassigned ("error") variation. Note that interactions in a particular design can be multiple, depending on the design.

For example, in a design involving two main effects and order, four interactions are possible: (1) main effect 1 and main effect 2; (2) main effect 1 and order; (3) main effect 2 and order; and (4) what is called a "second-order interaction," main effect 1, main effect 2, and order.

More complicated forms of ANOVA are possible, such as three-way analysis, three-way with replications, and so forth. With today's computational resources, such problems are less formidable than they once were.

However, recognize that the order of complexity of such problems grows rapidly, and the difficulties in interpreting the results grow accordingly.

Few people can think their way through interactions beyond the second order, for example, whether or not they can set up the correct sums of squares, degrees of freedom, and computational routines. The appropriate references may be consulted in case the reader is interested in delving further into this area.

VI. Covariance Analysis

In the section titled "correlative analysis," in Chapter 7, I described the idea of covariance analysis in the course of my brief discussion of partial correlation. At that time I gave a simple example, and pointed out that the control of the influence of an unwanted variable upon the relationship between two other variables of interest could be approached crudely through a categorical analysis or through a statistical model. (The reader may wish to glance back at the commentary presented in Chapter 7.)

Now that we have discussed ANOVA, it is probably time to mention that analysis of covariance (ANCOVA) finds it most frequent application in collaboration with ANOVA in a statistical model for the study of mean experimental effects.

Covariance analysis is really a branch of ANOVA, in that it is an extension of the analysis of the covariation in the data derived from a study. Practically, it is a set of procedures that are sometimes used in conjunction with ANOVA to control statistically the effect of unwanted variables on the comparison of means.

Suppose we have reason to suspect that there is an interaction between some variable (frequently a background sort of variable that we cannot easily directly control in a study) and the study variable of interest. For example, if our study variable is reading instructional method, but our student sample for the study appears to be rather diverse with respect to "IQ" (or scholastic aptitude, if you prefer), we might fear that the comparison of reading instruction means would be somewhat contaminated by differences in IQ.

As pointed out previously, a sort of "categorical" covariance might be attempted by sorting out the students into several levels of IQ (say, high, medium, and low), and conducting our study of reading methods separately within each group. (This would be a crude way to do what we can do statistically with analysis of covariance.) However, such a procedure breaks down rapidly as the problem is expanded to include more factors, replications, orders, and the like.

It is not necessary to go into the details of the covariance technique here. Suffice it to say that, in addition to the measurements of interest, such as reading performance by instructional method, one must have the value of the "covariate" measure (IQ in our example) for each of the subjects. In effect then the sums of squares in the ANOVA are "corrected" before the computation of the F test in such a way as to equalize the reading groups with respect to their levels on the covariate variable.

This is accomplished through the analysis of the cross-products in a manner analogous to the analysis of the sums of squares. (Here again we see the value of having variance and covariance terms of similarity and consistency in definition.) The net effect of the analysis is to enable the comparison of the group means on reading performance as they would have been had there been no relationship between IQ and reading performance. (Thus, it can be noted that ANCOVA accomplishes in studies of means essentially what partial correlation, mentioned in Chapter 7, accomplishes in studies of relationships.)

A wide variety of potential applications of the covariance analysis technique exists. It is not nearly as widely utilized, however, as is ANOVA. Part of the reason, perhaps, is that, just like regression techniques, the statistical equalization with respect to the covariate variable tends to capitalize on chance fluctuations and errors in the data set, and thus may be subject to some instability. In addition, it may be difficult to meet the various mathematical assumptions entailed.

VII. Point Estimates and "Confidence"

The last topic that we will deal with under the rubric of inferential statistics is that of point estimates and confidence. Until now we have been concerned primarily with problems that required us to contrast a hypothesis with the results of sample data. There are, however, situations in which we have no real hypothesis, but simply wish to estimate a statistical parameter for a population, based upon sample data. In such cases, we are looking for "point" (single value) estimates of population values.

Suppose, for example, in the case of the shoes discussed above, instead of wishing to test the manufacturer's advertising claim, I had simply wanted to estimate the mean wear time for his shoes. To do this I would take a sample of his product and submit it to wear testing. Then I would compute the mean wear time. Because it can be shown that the best estimate of the mean of the population is the mean of the sample, I would simply offer my sample mean as a "point" estimate of the population mean.

Such problems are very common. Whenever we take a sample of anything, compute a statistic, and generalize that statistical value to the population, we are making a point estimate.

A. Desirable Properties of Estimates

Such estimates should have a couple of characteristics, insofar as possible. They should be "unbiased." This means that the "expected value" of the statistic should be the population value. Divested of the statistical jargon, this is equivalent to saying: if you took an unlimited number of samples and computed the statistic, and then took the mean of all of those sample statistics, that mean would be the same as the population value for that statistic.

It turns out that the sample arithmetic mean is an unbiased estimate of the population arithmetic mean. Likewise, a sample proportion is an unbiased estimate of a population proportion. However, V as we have defined it is *not* an unbiased estimate of the population variance. In order to make it so, we have to substitute $N - 1$ for N in the denominator.

(The mathematical justification for this is beyond our needs here, but pertains to the issue of degrees of freedom.) Thus, when making a point estimate for the variance, we want to use $\Sigma d^2/N - 1$ for the variance. (Of course, when samples are large, it makes little numerical difference in the value obtained, so this is mostly a theoretical consideration.)

Another desirable characteristic of a point estimate is that it be an "efficient" estimate. This simply means that the SD of the estimate itself is the smallest possible for the population involved. In other words it estimates the population value with the least amount of error. Obviously this varies with population, but in normal and near-normal populations, the sample proportion, mean, and variance are the efficient estimates of their respective population values.

B. Confidence

In point estimation, as in other applications of inferential statistics, we must ask the question of how good is the estimate. Once again, we turn to probability statements for the answer.

As we have noted before, we cannot fashion a probability statement for a single value. Our known probability distributions represent probability as an area between two baseline values of the statistic being distributed. To address the topic of how good a point estimate is, then, we must turn the point estimate into a range! Let me summarize what happens, and then we'll review how we got there.

We construct a range around the point estimate in such a way that there is a known probability that the true, population, value lies in that range. The range around the point estimate is called a "confidence interval" because we have "confidence" equivalent to the probability represented by the range that we have bracketed the true value within the range.

This amounts to saying that in a probability distribution of the statistic, centered on our point estimate, there is the indicated probability that values that lie within the range can differ from the point estimate by that much by chance factors alone, without discrediting our belief that the point estimate is true. Conversely, by saying that the true value lies within the range, we would be wrong with the probability that a value lies outside of such a range if the point estimate were true.

There are two problems here. One is how do you construct a confidence interval, and the other is how do you interpret data that are reported in this form. I will go through the construction of a confidence interval for the arithmetic mean, but it is more important that you know what is meant when you find such a statement in the report you are reading. First, the construction.

1. Constructing a confidence interval

What we are looking for is a range such as this: $A < M_{pop} < B$. This says that the true, unknown, population mean lies between two values. *Note that the population mean is NOT a variable, and thus cannot be distributed.* Obviously, the two values, A and B, must arise from computations based upon the sample, specifically, the sample mean, because it is our best estimate of the population value. Thus, it is A and B that we must determine in order to have our confidence interval.

The first step is to decide how confident we want to be. Suppose we are only willing to be wrong 1 time out of 10 when we say that our interval really does include the population value. We want to have 90% confidence.

(However, because we could be wrong in either direction, we must divide our permissible error in such a way as to allow for this, i.e., 5% on either side, viz., to each of the two limits, A and B.)

Now, we are going to use the probability distribution appropriate for the statistic we are working with to determine how much range we have to allow on either side of a sample result in order to be sure to catch the population mean 90% of the time. Here I will use the normal distribution. (Technically, we can use the normal distribution only if we have taken a relatively large sample upon which to base our mean—say, at least 30 cases. Otherwise, we might use the "t" distribution.)

If we center our distribution on the sample mean (in effect, assume that it *is* the population mean), we want to find a point A on the left hand side of the baseline such that the probability of a value more negative than that is 5%, and a point B on the right hand side such that the probability of a point larger than that is 5% (5% to each because we divided our 10% permissible error, half to each side).

We remember that the normal probability curve arises from the distribution of a variable comprised of a difference divided by the SD of that difference: $(X - M_x)/\mathrm{SD}_{\mathrm{diff}}$. However, instead of being X, our variable in this problem is the sample mean itself. (What we are saying is that if we took an unlimited number of samples, we want to look at the distribution of those sample means.)

So, our sample mean is the "X" in the normal variate above, and the M is really the population mean itself. As I have previously noted, the standard deviation of the mean of a variable (x), for substantial samples, is $\mathrm{SD}_x/\sqrt{N}$.

Now, under the assumption that the sample mean *is* the population mean, and the center of the distribution, point A is equal to the sample mean minus the distance from that sample mean to the point on the baseline where there is a 5% probability of values more extreme. This point is read from the tables (which, remember, give you the area between the middle and the various baseline points) at probability (area) = .4500. So, the baseline value associated with point A is -1.645 (don't forget the minus, as it is on the left side).

In order to get this back into the units of the original scale, you must take what we were distributing in the first place: $(M_x - M_{\mathrm{pop}})/(\mathrm{SD}_x/\sqrt{N})$ and set it equal to -1.645. Remember that in the normal deviate distribution the mean is zero, so that M_{pop} becomes zero when we solve this expression. Solving this shows that point A is $(\mathrm{SD}_x/\sqrt{N})(-1.645)$ from the sample mean. Or,

$$A = M_x - (1.645)(\mathrm{SD}_x/\sqrt{N}),$$

in terms of the original X score scale. Analogously,

$$B = M_x + (1.645)(\mathrm{SD}_x/\sqrt{N}).$$

A couple of points need to be made. The first is that we have assumed a symmetrical range around the sample statistic. This is probably acceptable in most instances. However, even though they are not usually calculated, the possibility of asymmetrical confidence intervals does exist.

The second is that most people find it hard to remember that *it is the end points of the range, A and B, that are variables, not the population mean*. This is why we talk about the range bracketing the true value rather than the true value falling into a set range. It is the range which is a variable, based upon the sample observations!

2. Interpreting the confidence interval

Let's briefly discuss interpretation. First, using real data, you will probably be surprised at how broad the range around a sample value must be to give you high levels of confidence. To make a 99% confidence statement, it may be necessary to take a range so high as to be virtually useless. For example, if I were estimating someones age, I could probably say between 2 and 102 with virtual certainty—but what good would this do?

On the other hand, it is always possible to narrow the range. The precision of most statistics is a function of the number of observations. Therefore, one way to narrow the range is to take lots more observations into the sample used to compute it. For example, in the case above, it can be seen that the location of points A and B is a function of the probability required for the given confidence level and $\sqrt{N}$. Taking larger numbers of cases results in smaller ranges around the sample value for the same confidence level.

The other side of this relationship between the size of the confidence interval and the number of cases studied suggests that confidence statements may be used to estimate the number of cases needed in a study to give a specified level of confidence in the result. If you have decided what level of confidence you need for a study, and you can estimate the SD, you can work the confidence statement calculation backwards to see how many cases you need (what size N), to enjoy that level of confidence.

However, this analysis can sometimes reveal that the number of cases necessary to narrow the range to what is believed necessary may be beyond the resources available. Then the researcher must choose between accepting a higher level of error, more consistent with his resources, or deciding that the study is not worth the investment of resources in the first place given the expected imprecision.

From just a mechanical viewpoint, when you find a statement such as the mean is 22–24, with confidence of 90%, the proper interpretation is that there are 9 chances out of 10 that the population means really lies in this range. *However, you must understand that the probability statement refers to the whole range; a population mean of 23 is no more likely than 22 or 24 or anywhere between.*

Frequently, we find test scores reported with the score and a "standard error." Such data may be interpreted in a manner similar to the interpretation of confidence intervals. In effect the standard error is one standard deviation on the normal curve. So, this statement means that the reported test score ±1 SE will bracket the unknown true score about 2 times out of 3. (This statement assumes under the null hypothesis that the distribution is centered on the reported score, and that the errors of measurement are distributed normally.)

This is because, if you check out the tables, the probability value at 1.0 on the baseline for the normal curve is .34, and 34% on either side of the mean is 68%, or about 2 out of 3! Again, the reported test score, plus or minus one standard error, may be expected to bracket the unknown "true score" with a probability of about 2 out of 3. Thus, standard errors provide a valuable way, similar to confidence statements, to interpret the point estimates represented by test scores.

VIII. Summary

In this chapter I have tried to provide the basic notions involved in the second major branch of statistics—statistical inference. I have noted that inferential issues come into play in those cases where we wish to infer a statistical value or condition characteristic of a population from data based on less than the complete population.

Because it is impossible to prove something about a population based on only a sample of it (the critical cases may lie in the unexamined portion), the statistical logic of hypothesis testing involves setting up a null hypothesis, contradictory to the hypothesis of interest, and attempting to discredit the null. If the null is discredited, indirect support for the hypothesis of interest is obtained.

In testing hypotheses there are two kinds of possible error: rejecting the null hypothesis when it is really true (Type I or alpha error); and, accepting the null hypothesis when it is really false (Type II or beta error). Most statistical procedures deal only with Type I error, which is also called the "level of significance," because rejecting the null represents a significant support of the hypothesis of interest.

In an effort to control these various errors in making statistical statements, statisticians turn to probability statements to express the precision of their results. This is accomplished by contriving statistics with known probability distributions, so that the results achieved can be compared to what might have been expected under the condition of the null hypothesis.

There are several known probability distributions that are used for such purposes: the normal distribution, the "t," the "F," the chi-square, and the binomial are among the most frequently used. The process involves converting some statistic calculated from data collected into a derived form having one of the known distributions and then assessing the probability of such a result.

Examples of hypothesis testing were given, including the logic or one- and two-tailed tests of significance. It was emphasized that in cases where the null cannot be rejected on the basis of the data, it must not be considered as proved (no statistical hypothesis is susceptible of proof). Because the study that provided the data may not have been optimal in some way, or large enough, the same proposition may be supported by a further study.

Data should not be reused, however, because most of the statistical tests assume independence of data, and, indeed, reuse invites the capitalization on whatever errors and oddities might have existed in the data set.

Considerable time was spent on the subject of degrees of freedom, because many inferential tests depend on this concept. In effect, it reflects the extent to which the data being used represent independent facts or observations. Degrees of freedom are reduced by any constraints which operate on the data set in such a way to make one or more of the data points dependent (determinable from the others). An example is the limitation with respect to a mean that the sum of the deviations must add to zero.

It was noted that a technique called analysis of variance was able to handle a great variety of situations in which we are asking whether various means are different, and is capable of treating multiple means simultaneously. A simple problem was worked and some examples were discussed. A companion technique called analysis of covariance was also discussed briefly. This technique permits statistical adjustment of the means to be compared in analysis of variance to equalize for other factors that may influence the comparison.

Finally, I discussed point estimation and confidence intervals. In cases where we simply wish to estimate a population statistic based on a sample value, this is called a point estimate. To assess its goodness, we develop a range within which we can feel "confident" that the true value lies, based on combining our sample estimate with probability data about the assumed distribution of a variable derived from the sample statistic.

The important point is that it is the confidence interval itself that is a variable, not the population statistic. The correct interpretation of a confidence interval is that the indicated population value will be bracketed by the interval based on the sample with a stated probability (the confidence level).

It was also noted that standardized test scores, such as those from many published achievement tests, are frequently reported along with standard errors. The interpretation of a standard error is similar to that of a confidence interval. Generally, a test score plus or minus its standard error provides a range which will bracket the unknown "true score" with a probability of 2 out of 3.

IX. References

Agresti, Alan, *Categorical Data Analysis*. New York: Wiley, 1990.

Dwyer, James H. *Statistical Models for the Social and Behavioral Sciences*. New York: Oxford University Press, 1983.

Edwards, Allen L. *Multiple Regression and the Analysis of Variance and Covariance*, 2nd Ed. New York: W. H. Freeman, 1985.

Estes, William K. *Statistical Models in Behavioral Research*. Hillsdale, NJ: L. Erlbaum, 1991.

Flury, Bernhard, and Riedwyl, Hans. *Multivariate Statistics: A Practical Approach*. New York: Chapman & Hall, 1988.

Golberg, M. J. *An Introduction to Probability Theory with Statistical Applications*. New York: Plenum Press, 1984.

Loether, H. J. and McTavish, D. G. *Descriptive and Inferential Statistics: An Introduction*, 2nd Ed. Boston: Allyn and Bacon, 1980.

Neave, Henry R. *Elementary Statistics Tables*. London: Allen & Unwin, 1981.

Neter, John. *Applied Linear Statistical Models: Regression, Analysis of Variance, and Experimental Design*. Homewood, IL: Irwin, 1990.

IV

Statistical Surveys

For those of you who are still with me, I thought it would be useful to devote a chapter in this book to an "application" of statistics. By now you have gathered that statistics — or at least statistical notions and concepts — have considerable reach in today's world. Of all of the many arenas in which such ideas may be applicable I chose the statistical survey as being the one that more people of more differing interests might most frequently come in contact with.

Therefore, in the final substantive chapter of this book, Chapter 9, I want to walk you through the process of the statistical survey. Statistical surveys are done and reported on in this country in great numbers, and no person can be an informed consumer of such information without some background in how such surveys are done, and what sorts of conditions have a bearing on the outcomes.

It is certainly clear that statistical surveys are only one of many possible statistical applications. Certainly, some of you would prefer that I deal with other statistical applications, perhaps demographics, or quality control, or educational research, or measurement, or some of the many other applications areas. However, we have only the space (and the tolerance, I suspect) to examine one in some detail, and surveys have such broad impact on all of us that they seem the logical choice.

In looking at the daily paper, one sees that hardly a day goes by that some survey result or other is not presented to the public in the newspaper, or on TV, or in magazines. Such surveys treat a variety of topics from the mundane to the highly controversial. For example, recently I read in a newspaper that a survey had been done that showed that, contrary to what might be expected intuitively, more men than women of age 35 – 54 supported abortion on demand. I cannot speak for the accuracy of this survey, but certainly this is a subject of intense interest and debate these days!

I will not catalogue all of the surveys that are done regularly, and all of the topics that have been subjected to this type of investigation from time to time. Rather, let us proceed to review the survey process.

9

Statistical Surveys

I. Overview

Presumably each of the many "facts" to which the American public is
exposed daily has a basis in some information producing activity. Surely

surveys are among the most important of those information producing activities. Among surveys are found the basis for some advertising claims (" ... and, 80% of ____ trucks built in the last 10 years are still on the road;" election polls (" ... and, among Floridians, J____ C____ is leading his gubernatorial opponent by 5 percentage points with 17% undecided ... "); political polls ("the President's approval rating this week was an unprecedented __% ... "); and informational surveys of all kinds (" ... and __% of elementary school teachers have had at least one career shift in the past four years; or, nationally, the percentage of teachers with temporary teaching certificates is __%; or, the national percentage of once-divorced women who remarry is __%"; etc.).

In this chapter we will take a look at a variety of issues pertaining to surveys. We will see what constitutes a survey; we'll review sampling issues; we'll look at some of the more knotty problems associated with surveys; and suggest some of the cautions a survey consumer should keep in mind. This is followed by a review of what a survey report should be like. Finally, and most importantly, I want you to be aware of the limitations on survey information. The chapter closes with a detailed review of the specific steps necessary to design and implement a particular survey.

II. Surveys and Their Types

There is not complete consensus on the matter, but we shall use the term "survey" as a broad, generic term, defining it as follows: a survey is a collection of data from a large number of respondents (usually physically distant from one another) through the use of questionnaires, interviews, or some combination of the two. (This definition implies that the respondents are people rather than things. Collections of data about things are perhaps better described as studies, rather than surveys.)

The term may be used as a verb, meaning to conduct a study of this nature; or, it may be used as a noun to refer to the study generically, or to the process by which the data are collected, or to the results of the study, particularly when these are presented together. Naturally it is important to remain clear as to which of these related usages is intended in any particular instance.

Surveys, then, by definition, are not conducted in laboratories; they imply larger numbers of more widely dispersed respondents. They could be inferential in nature, in the sense that if one had a hypothesis involving a large number of people one might use a survey process to collect the data to test the hypothesis. However, in many cases they are descriptive, intended to reveal the way things are rather than to test a preconceived notion. This comports well with the underlying lexical definition of survey, which is literally to "look over." [Thus, the popular song of a few years ago entitled, "Hey, Look Me Over," might have been called, "Hey, Survey Me!"]

A. Types of Surveys

There are several distinctions which are often made in popular usage of the term survey, though there is frequently some overlap so that they are not precise categories.

The Census: The distinctive characteristic of a census is simply that the survey includes every member of the population, as opposed to merely a sample or subset of the population. (Censuses originated in early Rome, with the goal of having everyone registered for tax purposes.) In other respects it differs little from most other surveys.

The Sample Survey: This is by far the more common form. The subject matter may be anything, but the distinguishing characteristic is simply that the data are collected on a subset, or sample, of the population of interest.

The Poll: A subcategory of the sample survey is the "poll," since polls are almost never conducted on entire populations and thus are not censal in nature. This type of survey gets its name from its similarity to polling (voting). In effect, people are asked to "vote" on something, that is, to express an opinion. Hence, "public opinion" surveys are often called polls, and election-related surveys asking how people voted ("exit polls"), or plan to vote, are also called polls.

I should note that the methodological issues inherent in the conduct, processing, and reporting of surveys of various types are similar, with the exception of sampling, which of course applies only to sample surveys. The remainder of our discussion will focus on sample surveys, as these are the most general of these studies. Thus, I will not attempt to discuss censuses and polls separately.

B. Survey Requirements

Setting aside the issue of sampling for the moment, the requirements for a good survey are similar to the requirements for a good study of any kind. Care must be taken to define the issues, develop appropriate questions to shed light on the issues, apply appropriate analytic techniques, and interpret and present the results fairly and intelligently in the light of available knowledge of the field.

Although concerns with sampling are not exclusive to sample surveys, sampling is such a consistent focus of these investigations that the sampling process becomes the key factor that sets sample surveys apart from other sorts of studies. Thus, although I will try not to become too technical, it is necessary to spend some time on this crucial area.

III. Some Sampling Considerations

We have touched on sampling previously in Chapter 2. Any sample survey starts with a "population" of interest. This is the complete, well-defined group about which we wish to make some statements as a result of the study.

You will remember that sometimes it is impractical to study every member of a population of interest. In the area of surveys this is usually because it is too expensive, or too unwieldy, or too time-consuming, or, in some cases, not necessary for the needed level of precision. In any case, a sample is simply a subset of the population, chosen for the study. A very large proportion of surveys are of the sampling type.

A. Representativeness and Generalizability

However, to be useful for study in a sample survey, *the sample must be "representative."* In fact, all of the theory, and all of the various (and sometimes torturous) manipulations that samplers go through, are ultimately directed to this end!

Exactly what is a representative sample? Think of the population of interest, and of all of the aspects or characteristics of that population that may be important or may have a bearing on the issue under study in your survey. An ideal sample of this population would be a subset of the population containing, in proportion, all of these same aspects and characteristics as the full population. In other words, *to be representative the sample should be a small-scale replica of the population in all essential respects.*

Note that I have said "all essential respects." It's not important if the sample differs from the population in ways that don't have any bearing on the subject matter of the survey (if we can be sure that this is true). For example, if I am surveying to determine extent of functional literacy, I probably don't care if my sample differs from the population in terms of their choice of favorite color. However, I would be concerned about differences in percentage of males vs. females, or family income, or years of schooling, since these variables might well be related to functional literacy.

Why the extreme stress on representativeness? The simple answer is that without it there is no "generalizability." The usual purpose for studying a sample is to conduct a study of manageable size *with a view to generalizing the sample results to the specified larger population of which the sample is a part.*

We rarely care very much about the sample results for themselves, only about the inferences we can draw from them about the larger group. To the extent to which the sample is "biased," or nonrepresentative, we will be misled in the conclusions we draw about the study population. This is why we wish to have a sample that is representative of the population—so that the results from it can be generalized to that population.

B. Random Process

How does one "draw" (select) a sample so that it is a miniature of the population in all essential respects? Aren't there many variables that might be related to the subject of the survey? Yes, in most cases; and worse, some of these related variables might be ones you haven't thought of yet.

But there is one great principle in sampling—and most sampling methodology is simply a variation or elaboration of that one principle: To obtain a representative sample, select each member of the sample from the population according to a *random* process.

The randomness argument is that, if selection is left to random process (sometimes termed "chance"), each aspect of the population, no matter what, will be represented in the sample in proportion to its prevalence in the population from which you draw the sample.

It is important to understand what is meant by "random." Random means that each selection that we make for the sample is a function of chance factors, and is totally independent of (not influenced in any way) by any other selection. Another way of saying this is that each member of the population has the same, equal, chance of being picked for the sample.

However, there is a difficulty. Probably the least understood aspect of the concept of random sampling is that *random process does NOT guarantee that any particular sample is representative.* Random process only allows the laws of chance to work in distributing the population characteristics to the sample.

What is really true is that if we did it this way innumerable times, and averaged all of these innumerable samples, the *average* sample would be representative of the population.

Obviously, we rarely do the sampling more than once. Also, obviously, once is not innumerable nor an average. Therefore, it follows that any specific sample, even though drawn by random process, may be far from representative of its population. The distortion in the results of the survey that arises from this fact is called "sampling error."

In spite of this difficulty, random process is the primary mechanism statisticians use to achieve representativeness. Although there are many ways of defining and drawing a sample, almost all of them have randomness, or an approximation thereof, as their basis.

There are several implications of this problem of representativeness. First, survey users must realize that there is *always* sampling error in the reported results, and that estimates of the size of this error must be made and taken into consideration. Second, the characteristics of the sample should be checked against those of the population in whatever ways possible to determine whether the sample is highly aberrant or generally in the ballpark.

Finally, sampling techniques must be considered that improve the precision of the results beyond those obtained from the simple random sampling described above (and there are some such techniques available).

C. Some Sampling Types

We will talk more about the various kinds of error that can occur later. First, however, let's describe several of the sampling variations that frequently occur.

1. Simple random sampling

This case is the one suggested above, where the sample is drawn at random from the population with no preliminary manipulations to increase its efficiency. To implement this type of sampling you have to be able to identify each member of the population. This is required to ensure that each member of the population has equal probability of being drawn. This usually means a list of the population. Once a list of the population members is available, the individual members of that population are numbered, generally consecutively, so that each member has a single, unique numerical identification. Then the required number of population members is picked individually for inclusion in the sample, using some random process.

It should be noted that, although having your grandmother draw numbered slips from a hat is an approximation of a random process, good scientific practice requires that the survey sample be drawn using some *mechanical means* to ensure that the process is truly random. This is because humans no matter how hard they try, are rarely disinterested or unbiased about anything—sometimes without even being aware of their biases—and these biases can affect the selection of the sample.

Thus, the sample selection is usually accomplished by using a computer as a random number generator, or by using a table of random numbers (which was itself generated by a computer or some other mechanical means). Since not all of you may have a computer handy (heresy!), but everyone has access to some library where a table of random numbers may be found, let me go through the process of using a table of random numbers for you.

a. Using a table of random numbers. Suppose that we have defined, identified, and uniquely numbered a population of 5000 persons for our survey. The identification numbers run from 0001 to 5000, inclusive. On one or more of several possible grounds we have decided that a 5% sample, or 250 persons, will be adequate for our survey. Our task now is to randomly select 250 persons from our population, using a table of random numbers.

A table of random numbers is comprised of a bunch of pages covered with digits. Table 9-1 shows such a page. The critical characteristic of such a table is that each digit is independent of, or random with respect to, any other digit in the table, reading either up and down the columns or across the rows. This characteristic holds whether the digits are taken individually, or in groups or sets.

The first step is to achieve a random entry into the table. This is not critical, but you can easily see that it would be undesirable to start in the same place every time. One way to do this is to shut your eyes, open a phone book, and point to a number. The first digits of the phone number thus selected can be used to identify the page of the table of random numbers upon which to start. Then this process can be repeated to get the starting row on that page, and again to get the starting column.

Having identified the starting place, and having previously decided whether to read down or across from that place, you systematically read the table by groups of digits. A group of digits is comprised of the number of digits in your population size. In our case, where the population takes four digits (5000), we would read the table in sets of four. During this reading, every time we come to a number that can be found in our population list (i.e., any number from 0001 to 5000), that individual becomes a member of the survey sample. We continue with this process until we have accumulated the required number of cases—here, 250.

Table 9-1. Table of random numbers

67238601	91921192	57751963	30852995	95885402	99466607	99588340	05621041
78680532	17759116	74546852	67025185	45780867	66187372	49716175	00137253
56600865	73696987	72330630	20139621	43409979	09472908	30408611	30400165
69871478	85203820	55676167	13847997	65460169	83616084	92213246	24837473
21696527	92964699	79260090	73687994	61134737	52168804	21621285	21087078
15042130	36279205	46104975	24544218	70707541	36965440	67434722	04083024
74316056	17709863	12071532	19359136	00327681	48396275	73383529	75391434
23297409	48980572	96600793	99094014	53074192	96930935	16992846	53796452
19316322	78125046	99706300	83939852	16260724	85891832	54700164	07136172
43683640	97885251	25216294	95778328	04990236	85477597	73222517	15388584
48370606	35359844	10265898	11892727	61959682	88036519	47173667	65007509
58938554	22902606	61039308	49928886	13514695	44589983	02931293	74508531
33718592	50270886	94761981	84106127	81954280	82605908	88754174	51304944
79241290	98024031	87942821	92678197	54085496	05022183	51149384	79793682
21916371	52171831	30866365	36644202	06028149	70520105	17489475	40691260
35962393	00729517	88334974	69709548	84269473	55047817	84879662	61619709
27586441	07583370	66661019	79496879	42968824	94593745	15275016	98942517
75413813	77454822	86909067	60007010	37512365	09210037	36847649	50850563
27167706	88983227	64388070	15329869	74317750	08285997	51603900	01698058
52598907	18907133	94393553	94605433	16888817	77071527	49251527	97588480
52665476	43556153	44199801	58234664	48120274	44450559	27316888	53210509
24962438	22249401	51302044	75340450	88852384	31964500	17512133	89923973
85401468	61943933	27908650	52478538	72675653	84085661	18352393	05095776
46140576	63133179	71683306	93598474	76985522	79740702	34022962	72685840
49763802	46995385	82737213	91254543	10287048	25084065	19496489	75083044
59710896	63792047	86871264	49066701	55749596	27521411	31900164	66191158
06957010	82516354	65068241	35431516	71011526	50983062	88132903	65764783
92942381	61791618	89234996	68483106	44234879	52447434	28732543	84995286
48752025	02019719	64194642	32304073	06410063	99714476	87993020	83346741
50545418	78779536	75928739	24386446	99910546	70429110	70331563	56641864
91236189	57475389	46069480	34342613	03297538	32354664	32905879	56397951
08421802	89235476	92641884	53866261	12374681	04896311	74481342	40412814
68563249	88060307	97055975	17943314	77492738	01874509	30548178	54600723
07414733	89221145	77348979	77312870	97104275	55548437	74688656	46078338
26703359	88908444	67677504	02178137	67568355	21889431	26194270	07500651
47058819	65130281	04059731	17167373	89205223	67104423	63932935	52646920
17193080	79860802	16367599	67544823	28600990	35409378	46466164	23256609
79330073	62438653	26568998	16671655	37162327	98052732	38416266	27115325
90885378	74215418	53219950	74716896	25921147	83588830	43083526	07390754
02395914	54454060	22206799	58618377	96589296	89401361	41313394	43218600

For example, suppose that the first 10 four-digit groupings we found in reading the table were: 9967, 0453, 2456, 3901, 6766, 5167, 4449, 8006, 8903, and 5990. Then, so far, we would have selected the individual members of the population with the following identification numbers for inclusion in the sample: 0453, 2456, 3901, and 4449. We would proceed in this fashion to make 246 additional selections.

All of this appears straightforward, enough, I suspect, but there is one fly in the ointment. What do you do if a selected number comes up a second time, as might occur during a random process? Suppose, for example, that on your 233rd draw you get 2456 again.

b. Sampling with or without replacement. There are two choices. The first would be to include the responses from that individual twice in processing the data resulting from your survey. This is called "sampling with replacement," and technically it is the thing to do. Why? Because unless you do, you will not have the same probability of drawing each member of the sample.

Consider that on the first draw there is an equal probability of drawing any member of the population; it is specifically 1/5000. However, if you do not allow that case to be eligible to be drawn again, the probability of any of the remaining members of the population being drawn is no longer 1/5000; on the second draw it is 1/4999, and on the third it is 1/4998, and so on. This would be in violation of our definition which requires that each member of the population have an equal probability of being in the sample.

The second course of action, is to sample without replacement, ignoring the small change in the probability of selection as the drawing proceeds. For simplicity and other reasons, this is the way that it is usually done, particularly when the population is large, so that the change in probability is quite small.

(Sampling, of course, is used in studies other than surveys. Sometimes in such studies the measurements being taken as data result in the destruction of the sampling unit. For example, if you are testing the dissolving time of Alka-Seltzer tablets, any given tablet can be tested only once. Here, if sampling with replacement is imperative, the only solution is to weight the individual result by adding it into the data set once for each time it was drawn in the sample.)

You should remember, however, that sampling without replacement is only an approximation of random sampling. Furthermore, it is a *good approximation only if population is large, and the size of the sample is small relative to the size of the population*. In our present example, it would usually be considered acceptable.

2. Systematic sampling

The procedure known as "systematic" sampling is not really a random procedure with respect to the selection of each and every member of the sample. In this case, instead of randomly selecting each case in the sample, we randomly select a starting point, and then systematically take every "nth" member of the population.

In this procedure it is again usually necessary to identify each member of the population to be sampled, as was the case with the simple random sample. The exception would be cases where the population to be studied was the result of an ongoing process. For example, as a quality control study, we might wish to study a sample of toasters from a production line. Here we might decide to define a sample as being every nth toaster produced during a selected work week. Such a procedure might then be repeated periodically as a method of monitoring quality.

The first step in systematic sampling is to determine the size of "n." To determine "n," we first decide how many cases we wish in the sample. Then we take that number and divide it into the total number of members of the population. The result gives us "n," which is a number such that when we list the population, taking every nth member will give us the sample size we want. For example, in our previous illustration we would divide 5000 by 250, which yields 20. Then, taking every 20th case will give us a sample of 250!

You will notice that so far there has been no random process at all. We have simply managed to spread the sample out with respect to the population list, ensuring that it does not clump up at any particular point in the list.

Randomization enters in only in selecting the starting point. Thus, instead of starting at 1, we take the first group of 20 and pick a random starting point within that group. *Then* we take every 20th case. The random starting point is usually picked with the aid of a table of random numbers.

A systematic sample is usually a fairly good approximation of a simple random sample. However, there are sometimes particular population characteristics that you especially want to see reflected in the sample, in order to ensure representativeness. Sometimes these can be manipulated by adjusting the order in which the population is listed before the sample is drawn.

For example, if it is desirable to sample school districts in such a way that geographic distribution is ensured, the population list may be prepared by listing districts within state and arranging the progression of

states so that the states are taken in a contiguous order. Such a scheme
forces the systematic process to pick districts in proportion to the number
of them in each state and to virtually assure (depending on the size of the
sampling group) that each state has some representation in the sample.

Sometimes it is possible to control more than one factor through this
process of arranging the population list before drawing the sample systematically. For example, in the above illustration, if the districts are listed in
order of size within each state, the process will also tend to control the
representativeness of this factor in the sample as well as geographic
distribution.

The fact that the ordering of the population list can influence the
characteristics of the sample can also be a disadvantage, particularly if you
become aware of it only after the fact. For example, if you have inadvertently arranged the population list so that an important factor is systematically omitted from the sample, there will be a problem. Also, there is
always the possibility that the drawing interval (n) will match up with some
other factor in the population, producing an unintended bias. As ever,
there is no substitute for knowing the study area thoroughly.

3. Purposive samples

I have noted that humans do not operate randomly. Thus, most sampling
schemes feature some mechanistic principle to ensure randomness. However, in a few instances, notably ones where it is deemed highly important
to absolutely ensure the inclusion of particular characteristics or cases in
the sample, we abandon the principle of randomness.

Such samples are called "constructed" or "purposive" to call attention
to their departure from the more usual random process. An example
might be where you wish to study a rather rare phenomenon. Rather than
taking a sample by more usual means and running the risk that the
phenomenon of interest, or at least some variants of it, might not be
included in the sample, prime instances of the phenomenon and each of
its variations are deliberately picked out for inclusion.

The problem of rarity extends further. Suppose your sample contains
examples of the rare phenomenon you wish to be sure is included.
However, it contains so few of them that the data you collect about them
are based upon too few cases to be stable. Again, you might turn to a
purposive sample.

The use of purposive samples requires real *caution with respect to
generalizability*. Because the purposive sample differs from the population
in important ways—on purpose, as it were—we must be very careful

about attempting to generalize findings based on it to its parent population. Depending on the circumstances, the findings about the specially included cases might be reasonably representative *of those cases* (which is what we would hope).

However, in a larger sense the sample is *not* representative of the parent population because of the nature of its selection. Thus, *purposive samples tend to be oriented toward finding out something substantive about the nature of a phenomenon, and NOT toward establishing the prevalence of that phenomenon in a population* (*which is more a matter of generalization*).

It might be noted in passing that much medical research is dependent on this kind of sampling. This is because many medical conditions are too rare to become the subject of the more traditional sample survey, and their characteristics and the efficacy of their treatments must be studied on a purposive basis.

4. Stratified random samples

I have noted above that simple random samples leave the control of all factors of importance strictly to chance. In many cases this is neither desirable nor necessary. We will now discuss the most popular of the sampling strategies—stratified random sampling—with a view to finding out the various ways in which simple random sampling may be improved upon.

You will recall that the "sine qua non" of sampling is that the sample be representative of the population of interest. Random process is one way to gain representativeness. But, I reiterate, random process is dependable in ensuring representativeness only when repeated over large numbers of samples. Given any particular sample, randomly drawn, there is no guarantee that it will be representative of the population from which it is drawn in all essential respects. And, indeed, it may be differentially representative with respect to different variables!

This is clearly a difficulty. However, on the other hand, it is equally clear that, given the complexity of most real-life situations, one cannot expect to arrange for individual control of all of the possible variables of concern.

The answer that has found the most favor overall is, not surprisingly, a combination of deliberate control and random process; it is called the stratified random sample.

In the stratified random sample the population is separated into nonoverlapping subgroups that cover the entire population (mutually exclusive and mutually exhaustive groups). The separation is accomplished according to categories on a "stratification" variable which is thought to be important to control (assuring that its effects are for all practical purposes eliminated from the study variable). Then, a random sample is drawn *within* each stratum (subgroup).

(Some readers will note the similarity of this concept to the idea of categorical covariance analysis. Indeed, categorical covariance, ANCOVA, and partial correlation techniques are all also intended to introduce a means of controlling the effects of selected variables on study findings and comparisons under differing circumstances.)

For example, if we were to embark on a study of the distribution of education in the adult population of the United States, we might draw a random sample of the population and ask questions about the amount and kind of education each member of the sample had experienced. However, if by chance, our sample were not truly representative of the population with respect to its distribution on chronological age, our education information would probably be distorted.

This is because there is a clear dependence between age and amount and kind of education. Older people tend to have had less formal education than young adults, since education was less frequently required early in this century than it is now. Also, of course, younger people may not have completed all of the education they will eventually have.

You might set up age categories such as: 21–30, 31–40, 41–50, etc., and then draw a random sample within each of these categories. At the end of this procedure you could make clear statements about the amount and kind of education within each age group. If you wanted a national estimate, you would probably combine the results after weighting the result in each category by the relative proportion that that category bore to the total population.

Thus, if the median result in the 31–40 category were 14.6 years of schooling, and that category comprised 20% of the adult population, they would contribute 20/100ths times 14.6 to the national estimate. (More about weighting later.)

By doing your study this way, you would avoid the happenstance that too many older people might fall into the sample, just by chance, and bias the results toward the lower side—or vice versa. Also, you would be depending on random process to adequately distribute any other characteristics of the population into the sample in roughly the same proportions in which they existed in the population itself. The key point is that by using the stratification method, you don't take chances with the age variable.

Sometimes it is desirable, or necessary, to stratify with respect to more than one variable. For example, in the above illustration you might decide that you must have control of geographic distribution as well as age. This means that each age category must then be subdivided into geographic strata as well. If you took four such strata, then, combined with, say, seven age strata, you would have 4×7 or 28 separate strata to work with.

It is easy to see that the stratification process can quickly get quite out of hand. With just the two stratification variables suggested you are already dealing with 28 subpopulations. To get a representative random sample within each of the 28 implies a very large total sample. Adding another stratification variable would multiply the problem even further—that is, by the additional number of strata required. For this reason, stratification is usually limited to two or three variables and the number of categories on each variable is usually limited as much as possible—frequently to two or three or four.

Not just any variable can be used for stratification purposes. Stratifying on a variable that has no relationship to the variable under study does not improve the results over that derived from simple random sampling. This is because the stratification cuts across the study variable, producing strata that differ from each other with respect to the study variable only by chance (i.e., the stratification variable).

On the other hand, stratification on a highly correlated variable produces strata that are markedly different from each other with respect to the study variable, thus ensuring that the study variable can be examined in other respects *for each level of the stratification variable.* In effect, then, this removes the effects of the stratification variable (controls it) from all of the other study comparisons.

In fact, ideally, in a stratified sample the strata should be as different as possible from each other with respect to the study variable, but the members of the sample should be as similar to each other on this variable as possible *within* each stratum. (Stated statistically, the inter-stratum variance should be at a maximum, whereas the intra-stratum variance should be minimal.)

Now, let's sum up. The stratification variable(s) must be ones that you know are related to the variable(s) being studied. (Stratification on an unrelated variable is equivalent to no stratification at all!) Because the process exacerbates the complexity of the study rapidly, stratification should be attempted *only* with a very limited number of variables for which you feel that control is crucial.

Stratification enables you to guarantee that the sample will be representative with respect to the variables of stratification, while representativeness with respect to other variables is left to the workings of random process within the various strata.

Statisticians will say that stratification with respect to appropriate variables increases the "efficiency" of the sample. By this they mean that a given amount of information about the population can be achieved with a total sample of smaller size than would be required using simple random sampling.

D. Sampling Ratios

So far we have been discussing, by implication at any rate, what statisticians call "pps" ("probabilities proportional to size") sampling.

To illustrate, if the sampling stratum "A" contains 1000 items, and sampling stratum "B" contains 4000 items, pps would indicate that we draw one item from A for every four items from B. In this way the sample would contain the same ratio of A items and B items as does the population. Thus, if we wanted a total sample of 250, we should sample from the two strata in proportion to their relative size, or 20% from A and 80% from B, that is, 50 cases from A and 200 from B.

Another way to put this is to talk about an overall sampling ratio of, say, 5% or 1 in 20. Such a sampling ratio, if applied to A would yield .05 times 1000 or 50 cases, and if applied to B, .05 times 4000, or 200 cases—a total sample of 250 cases. Thus, the application of a single sampling ratio, here 1 in 20, to all strata, produces a sample in which the strata are represented in proportion to their size in the population.

There are times when this is neither desirable or appropriate. Suppose stratum A were comprised of large high schools, and stratum B of small high schools. We might easily decide that a sample of 50 large high schools would be insufficient to represent the rich variety of this type of American education. Yet at the same time, practical limitations of time and money might well limit the study effort to a total sample of no more than the 250 we have been talking about. If we feel that the schools in stratum B would be adequately represented by some smaller size of sample than 200, say 150, then we might decide not to sample according to "pps."

1. Differential sampling ratios

To get a sample of 100 from stratum A and 150 from stratum B, we would have to take 10% in stratum A and 3.75% in stratum B. This works out to 1 in 10 in stratum A, and approximately 1 in 26 in stratum B. (This will actually give us slightly more than the desired sample, that is, about 154; if we must stay under a total of 250, we will take 1 in 27, yielding 148 for stratum B.)

What we have now done is called "using differential sampling ratios." This procedure is used whenever it is thought that straight pps sampling will not yield enough cases in a stratum to adequately represent the stratum. This procedure is in quite common usage, particularly where the stratification variable used results in strata of widely disparate sizes, but it is still thought to be important that each be fully represented in the final, total sample.

There is a significant complication, however. If you stop to think, you will easily see that you can no longer simply combine the results from the various strata in order to get your overall, population estimates. The use of differential sampling ratios has distorted the distribution of the various characteristics of the population as estimated from the sample.

For example, in the illustration above, whatever you find out about high schools will be overly influenced by whatever characterizes big high schools, since when you drew your sample you took proportionately twice as many as really existed in the population. In order to compensate for the distortion introduced by the differential sampling ratios, you must "differentially weight" the results from each stratum when you combine the strata into the total sample.

2. Inflation

This is accomplished when the sample results are "inflated" to population estimates. In the first example, of pps sampling, where the sampling ratio was a straight 5% (or 1 in 20) sample values are inflated to population values by dividing by the sampling ratio (or, equivalently, multiplying by its inverse).

Thus, if the number of teachers in large, A schools were found to be 3500 in the sample, the estimated population number of such teachers would be $3500/.05 = 70,000$. Also, if the number of teachers found in the small, B schools were found to be 3000, the population estimate would be $3000/.05 = 60,000$; and thus, the total number of teachers would be estimated to be the sum of the strata estimates, or 130,000.

In the differential sampling example, however, the different sampling ratios must be used as the basis of the weights. Where we had a 10% sample in stratum A, we would divide the sample values by .1 (or multiply by 10), and where we had a 1 in 26 sample in stratum B, we would divide the sample values by 1/26 (or multiply by 26). Only then could we add the stratum estimates together to get an overall estimate.

In a nutshell, then, where the same sampling ratio is used throughout, you can add across strata and then inflate. But, where differential ratios are used, you must inflate the stratum values with the appropriate weight *first*, before adding together to get the final, overall estimates.

There are a number of variations on these themes, but the above should be sufficient to convey the general outlines of stratified random sampling.

E. Multilevel Sampling

In each of the methods of sampling discussed so far the individual members of the sample have been selected in a single draw. It is possible, however, to do a sequence of draws to get down to the final sample. There are two main types of such multilevel sampling, "cluster" sampling and "multistage" sampling. We will take a look at each of these.

1. Cluster sampling

Suppose we wish to take a sample of high school seniors in the state of Maryland. As you will have noted above, the most common methods of sampling require a list of the population from which to draw the sample. Such a list of all high school seniors in Maryland would be difficult to compile. However, we can easily get a list of all the senior high schools in Maryland from the State Education Department.

Suppose we take a random sample of high schools, and then define the study sample as "all high school seniors in the selected sample of high schools." This method of sampling high school seniors would be called a cluster sample, because the individual members of the sample were not each individually and independently selected, but rather were chosen in groups or clusters.

In the event that cluster sampling would produce too many cases in the final sample, which might arise from the need to select a sufficient number of schools to adequately represent the population of schools, the cluster sample can be reduced by randomly discarding the excess cases.

2. Multistage sampling

Suppose we first take a random sample of schools, then take a random sample of teachers from the selected schools, and then follow by taking a random sample of students taught by the selected teachers. This procedure would be called a "three-stage random sample."

The difference between this procedure and the cluster sample is that in this procedure we have managed to maintain an approximation of the idea that each person in the population should have the same chance of being selected. The probability of any particular student being selected for this sample is the probability of drawing his school, times the probability of drawing his teacher given that his school has been selected, times the probability of drawing him, given that his school and his teacher have been selected.

I say an approximation because the probabilities for different students in the population differ somewhat as a function of the differing sizes of the subpopulations at stages 2 and 3. That is, among those schools selected in stage 1 there will be differing numbers of teachers who meet the qualifications for inclusion in the sample; likewise, among those teachers selected there will be differing numbers of students from which the final sample is to be selected. These differences result in somewhat different probabilities of selection for different students.

But, the important thing is that the probabilities of inclusion can be estimated for each sampled individual. This makes possible the construction of weights that can be applied to the problem of estimating population values.

On the other hand, in the cluster sample process, once your cluster has been selected, with whatever probability, your chance of inclusion in the sample is 100%. One of the effects of this characteristic is that it restricts the variance as compared to independent selection of each individual sample member. This has an effect on the estimation of sampling error for such samples. In addition, while population values can still be estimated, the estimations are less precise.

Although the statistics of error estimation are greatly complicated by multilevel sampling techniques, these techniques are very popular. The major reasons for their popularity are not hard to understand.

In the first place, it is frequently difficult to come up with the population listing required for random, systematic, and stratified random sampling when the population is large, diverse, or otherwise inaccessible. Then, too, multilevel samples are sometimes more efficient because they can be tailored more completely to the exact needs of the study. Finally, they offer the possibility of a linkage among the various levels that is more direct than independent sampling of each level would be.

An example of this latter point follows. You might take a sample of schools and describe today's school curriculum, teacher training, physical plant, and the like. You might take a sample of students and describe their course-taking behavior, grades, and disciplinary problems. You would learn more, however, if the sample of students studied were attending the very same schools from which the school information had been collected. This "linkage" would make it possible to relate specific school characteristics to specific student behaviors. Such an investigation implies a multi-stage sample.

F. Panel Sampling

The final variation on sampling that I will mention is called panel sampling. In technical respects a panel sample is constructed through some combination of the methods outlined above. However, it is called a panel because, once constructed, it is used repeatedly in a manner similar to the use of an advisory panel. Panel samples are popular with pollsters, market researchers, social demographers, TV ratings organizations, and the like.

A panel is usually constructed for a study that is to be repeated several times. It tends to be a small sample as compared to the size of the population in order to control the costs to the polling organization. TV ratings have been done on panels as small as 1200 families. Because panels tend to be small, considerable efforts are usually undertaken to make them as efficient as possible.

In order to maximize efficiency, panel samplers compile large amounts of data from the census, from past studies, from demographic sources, and the like. Their selection of members of the panel carefully attempts to assure representativeness by using these various sources of data to ensure that the panel contains the appropriate mix of each of a number of factors that might bear on the area to be studied.

Thus, panels tend toward being purposive in nature, or at least to feature comparatively large numbers of controlled elements in their selection. Conversely, they tend to depend less on the vagaries of random process than do other types of samples.

Because this process of construction tends to be expensive, panels tend to be reused. To the extent possible, conclusions drawn from these samples are checked against reality, new data are applied, and their compositions are reanalyzed and revised.

In some cases panels are alternated, in order to avoid overburdening their members. In other cases, membership is constantly replaced by rotating in a few new members from time to time, so that the entire panel membership turns over in a specified period of time. In these ways panels tend to become more efficient and predictions based upon them more accurate.

These advantages, however, have corresponding drawbacks. To the extent that the panel members are biased, or not representative in important, and perhaps unknown, ways, their nonrepresentativeness is perpetuated from study to study.

In addition, repeated use of the same panel members runs the risk that the mere fact of participation can have an increasing impact over time on the response tendencies of the individual. Only the repeated drawing of entire new samples can prevent this. In addition, obviously, to the extent that the background data and historic information that facilitates their construction does not exist, or cannot be related to the study objectives, panels cannot be so effectively constituted.

There are many additional variations on these approaches to the sampling issue. It does not serve our purposes to attempt to be exhaustive here, so we shall go on to examine some of the other issues pertaining to surveys.

IV. Errors and Nonresponse

In a process as complex as surveying, there is much that can go wrong. We will skip the obvious errors that can be made by failing to define the problem, population, procedures, etc., adequately. We have mentioned those before. However, we should look at least briefly at the two main categories of "statistical" errors, and at the problem of "nonresponse" which can create a lot of error. I have previously mentioned "sampling errors," so let's start there.

A. Sampling Errors

In effect, when you use a value computed from a sample as an estimate of the corresponding value in the population (a population "parameter"), your estimate is inaccurate as a function of sampling error. This sampling error arises from the fact that every sample is an imperfect representation of the population (in spite of our best efforts to the contrary).

The size of the sampling error is a function of the type and nature of the sample, the size of the sample (other things equal, the larger the sample, the smaller the error), and the thing being estimated and its distribution. The exact computation can be very complex in some situations, and is beyond our interest here.

A particular example of the way sampling error applies to survey population estimates harks back to our discussion of confidence intervals in the preceding chapter. In the case of the confidence interval, we created a particular statistic with a known sampling distribution (i.e., distribution of sampling error), and used that information to construct an interval within which we would expect to find the population value with a stated probability.

Theoretically, we could construct such an interval for every population estimate, based on the distribution of its sampling error. Practically, such computations are tedious and often excessively involved. Frequently, approximations can be computed that serve a similar purpose.

You recall from our discussions of confidence intervals, if you start with some unknown population parameter that can be estimated from a sample, each possible sample would provide an estimate. Because successive samples would differ by chance, these estimates would also differ.

If you distribute these different estimates of the same population parameter, they will tend (in most cases) to center on the population value itself. As for any distribution, we can estimate its variance. It is this variance that is the error variance arising from sampling, or the "sampling error." The usual index of its size is its standard deviation, and this is called the "standard error of estimate" for this particular population parameter.

Obviously, if you knew the distribution you could easily calculate its standard deviation, and this would be the measure of the sampling error for this estimate. But, recall that the sampling distribution is what would occur if you took innumerable samples and averaged the results—and nobody is prepared to do this. Thus, the sampling distributions, and their standard deviations or sampling errors, must themselves be estimated, either mathematically or through approximations.

Practically speaking, survey results should be reported with estimates or approximations for the sampling errors. These estimates may be in the form of confidence intervals, where these are appropriate. They are frequently reported as "standard errors," which are estimates of the standard deviation of the distribution of hypothetical sampling estimates around the population value. They are sometimes in the form of "generalized variances" which are the same thing, but less precise, since they have been calculated for groups of population estimates rather than individually. However reported, some estimate of sampling error should be reported with every survey's results.

The interpretation of these reported estimates of sampling error follows the guidelines discussed in the previous chapter under confidence intervals and the standard errors of test scores. In general, the estimate of sampling error provides the basis for saying that *the reported sample value, plus or minus the error estimate, brackets the true (population) value of the statistic with a specified probability*. Obviously, the higher the probability, the wider the range will be, or conversely, the smaller the range required, the lower the probability that the true value will be bracketed.

Another, and much less satisfactory, way of reporting the precision of survey results is the use of the "coefficient of variation (CV)." The CV is simply the ratio of the standard deviation to the mean of the distribution. This ratio provides the "spread" in the distribution relative to its mean.

However, there is no intrinsic standard to which any given case can be compared. Generally, smaller is better—which simply says that the less variability there is the less likely it is that a large portion of it will arise from sampling errors. However, the measure is simplistic. Frequently an arbitrary standard is adopted, such as "CVs should be less than 0.1, or 10%."

B. Nonsampling Errors

In addition to having to take into account sampling errors, there are always considerations of nonsampling errors. Generally our approach to controlling the size of sampling errors devolves to increasing the sample size. But, there is no such simple solution to the control of nonsampling errors.

As is clear from the above discussion, the sampling errors generally arise from the vagaries of chance introduced by random sampling or some form of it. Nonsampling errors, on the other hand, arise from the introduction, in some fashion, of some *systematic bias* in the sample as compared to the population it is supposed to represent. Even worse, while some such biases are known or can be predicted, there is always the possibility that other, unknown, sources of systematic bias exist.

Let me try to elucidate with some examples:

1. Inappropriate definition of the population

In the 1936 presidential election, polls predicted a much stronger showing for the Republican candidate, Alf Landon, then actually occurred. Pollsters had taken a sample from a population of telephone directories, since comprehensive voter rolls were not available to them. In truth, a significant portion of the voter population did not own a phone at that time, and that portion of the population was heavily Democratic. Thus, the survey, being confined to telephone subscribers, was systematically biased toward a Republican outcome. Only after Franklin Roosevelt won overwhelmingly was it recognized that failure to include significant portions of the population in the sampling process had greatly influenced the accuracy of the polls.

2. Inappropriate questions

Let's suppose that you are taking a survey for "Krunchies." Your first question is, "Give three reasons why you like Krunchies." You are surprised to find that you have a much higher approval rating for Krunchies than the last surveyor got. You examine his questionnaire, which starts out with, "Do you like Krunchies?" You realize that yours was a "leading question" which started out with the implicit assumption that people were familiar with and liked Krunchies.

Such an implicit assumption is obvious to the respondent on an internal level, and tends to intimidate him or her. Not wishing to contradict the interviewer, many people will respond without troubling to inform him or her that they really didn't care much for this cereal in the first place. People do that, you know! (The Census Bureau has discovered that simply changing the order of the same words in a question can significantly alter responses.) Questions must be crafted to be as neutral to the results as possible to avoid bias.

3. Inappropriate content

Content may be biasing in many ways. If you write your questions in college level language, less educated respondents may either misunderstand your questions (with unpredictable results), or lose sympathy with, and quit trying to understand, your inquiry. If your content is offensive, or controversial, or simply uninteresting to one or more segments of the population, responses may be biased, or people may simply not answer at all.

4. Differential exposures

If, for any reason, your sample has differential exposure to the material being covered by the survey than does the population, a bias will exist. This is one of the problems of sampling for convenience. You want a sample, but you don't want to travel to get it. So you take respondents only from your own neighborhood with whatever nonrepresentativeness that entails.

Or, you are running an experiment in college psychology lab. You use your classmates as subjects. [It has been said that the entire body of psychological knowledge is suspect, since most of it was developed using psychology students as the subjects!] Or, you sample alphabetically, forgetting that some factors, such as income, may be related to alphabetical sequence in some populations (where, for example, certain names belong to later immigrant groups, which by and large, hold less remunerative jobs).

5. Inappropriate data handling and processing

Any incident in the handling and processing of the data that creates an imbalance in the results is a nonsampling error. For example, in a study in which I was a participant many years ago, a data processing clerk dropped a box of 2000 IBM cards, set them aside temporarily, and never got them into the data analysis. Since the cards were arranged by Census District, a whole geographic unit was omitted, biasing the results.

6. Summary

Nonsampling errors can arise from *any* factor that *systematically* (nonrandomly) predisposes the results of the survey in a direction misrepresentative of the population. There is no simple control for such nonsampling errors. Professionalism and care in the entire survey process, along with intelligence and deep knowledge and understanding of the field of inquiry, are the best safeguards in this area.

C. Nonresponse

After all of the trouble you have gone to in order to design a good survey —problem, forms, sample, procedures, etc., one hurdle still remains, one over which you have little or no control. Suppose those rascals out there don't respond. Suppose they don't agree with you that responding to this survey is important—or they think it takes too much time—or it gets lost in the press of other business. How valuable are your design and careful craftsmanship now?

This is the problem of "nonresponse." There are only two alternatives: report what you get as is, with appropriate caveats and qualifications, of course; or, "fudge"! Having invested a great deal of time and money (and pride) into the survey by this time, some people choose to fudge.

The choice of how to fudge (how to treat nonresponse) is rarely simple. In the first place, there are really two types of nonresponse, and most surveys must deal with both—although not necessarily in the same fashion. The first, and more serious, is instrument nonresponse. This is where the sampling unit (respondent) failed to respond at all, and you don't even have a form from him or her. The second is item nonresponse. This is where the respondent returned the form, but failed to answer one or more particular questions.

1. Instrument or form nonresponse

Let's look at form nonresponse first. As you might expect, the first question is how bad is it. Things are quite different if 95% of your sample responsed than they are if 55% responded, or only 25%. The smaller the response percentage, the greater the likelihood that those who responded are significantly unlike those who failed to—and the more trouble you are in.

In other words, a high level of nonresponse strongly suggests that those responses in hand are subject to systematic biases (nonsampling errors). Unless you are in a position, by recourse to other sources of information, to assess and evaluate the nature and extent of the impact of such biases on your survey results, you are on very thin ice when you try to generalize the sample results to the population.

How much is too much? A fairly widely accepted statistical rule of thumb calls for a return of 75–80%. That is, if your form nonresponse exceeds 20% or so, and you cannot make justifiable adjustments based on outside sources, your survey results must be regarded as unreliable.

There are exceptions, of course. One of them is based on what else is available. If, for example, there is already available, comparable information from other sources, there is less inclination to accept a new survey with a low response rate. If, on the other hand, there is a paucity of information in this area, a carefully caveated survey of relatively low response rate may provide some useful information.

It may even be desirable to publish the sample results as is, without attempting to inflate them to population estimates. In such a case, however, it is extremely important that the researcher reveal in detail the limitations of his or her study so that his or her results are not given the same weight that they would bear with an acceptable response rate.

If the importance of the study warrants it, and if resources are available for the purpose, a survey with a low response rate may sometimes be subjected to a "nonrespondent" study. Such a study attempts to gather a limited number of items of critical information from those who failed to respond to the main study. Frequently, these items are focused on describing the nonrespondents in terms of suspected areas of disparity from those who did respond to the survey. This sort of focus is adopted where there is the possibility of using such information to make statistical adjustments to the findings.

In a nonrespondent study a sample of nonrespondents is drawn and subjected to very intensive efforts at locating them and securing their cooperation. These results are then used in the main study to guide the assessment and evaluation of the nonsampling errors in the main study, and to make appropriate adjustments and corrections to the population estimates to compensate for such errors. Sometimes these corrections may call upon other sources of data, where the critical items facilitate the tie-in to the alternate data sources.

If the survey should be a follow-up to a previous, acceptably complete, survey or data collection, frequently the necessary caveats, corrections, and adjustments can be constructed from an analysis of the data which the nonrespondents had supplied in the earlier collection. In such cases, a nonrespondent study may not be necessary, because the same purpose is achieved by the analysis of the previous data.

In cases where the nonresponse level is moderate, the usual procedure is to "impute" for the nonresponse. I will discuss imputation shortly.

2. Item nonresponse

The other type of nonresponse, item nonresponse, usually does not imply discarding the survey. In most cases, the item can be discarded and the remainder of the survey utilized. Of course, if the item in question is critical to the study, there may be a question as to whether or not the remainder of the study is worth analysis and reporting. This must be judged on a case-by-case basis.

Frequently nonresponse on critical items is made the subject of intense follow-up procedures, and an effort is made to bring the response rate up to an acceptable level just for those items deemed to be critical. Sometimes it is possible to impute for missing item data (see below).

3. Imputation

Let's take the usual case, where the nonresponse rate (either form or item, but I will talk about form here) is generally acceptable—say 90%. As mentioned earlier, we can either report the results as is, or make some adjustments. In general, the study will have been designed to study a particular population in some respect. The sample of course is only a fraction of that population. Therefore, we don't usually report sample results as is; we usually employ them to construct population estimates.

In the process of doing this, it is common practice to factor in the missing respondents (in this example, the missing 10%) in some fashion. The purpose of doing this is so that we can talk about the population as a population rather than having to continually qualify the numerical estimates as being 90% of the population. Raising the response rate to 100% in some way involves imputing the results of the portion of the population that responded to the whole, intact population. Therefore, the process is called imputation.

All of the various imputation methods involve the substitution of data from some other source for the data that are missing in the survey. These methods may be variously successful in terms of accuracy in replacing what would have been the responses, if they had been received. How successful is usually unknown in any specific instance.

In addition, wherever the process results in the reuse of existing responses for missing ones, the standard deviation of that variable will be reduced, as compared to what it would have been if the missing responses had been available. This will have undesirable effects on confidence intervals, correlation coefficients, and other statistical indices.

a. Direct proration. There are several ways to accomplish imputation. The first, most common, and simplest is direct proration. For example, if you are imputing for the missing 10% in the paragraph above, and you had 9000 teachers based upon your 90% response, you would divide 9000 by 0.9 to get 10,000 teachers. The 10,000 teachers is what you would have if the missing 10% of the respondents had answered the same way as the 90% who responded did.

It is important to note that the direct proration procedure for imputation makes the assumption that those missing are exactly like those who responded. This is not likely to be strictly true, but nonetheless this method of imputation is the most widely used.

b. Matching. Other ways to impute include matching. In the matching approach you attempt to find a respondent whose known characteristics match those of the nonrespondent, and then double the weight assigned to that respondent. In effect, you substitute the matching respondent for the nonrespondent by adding his data in a second time.

Obviously, this method is dependent on knowing some critical characteristics about the nonrespondent upon which the match may be based. In some cases the variables used to construct the sampling plan are the ones used. In cases of item nonresponse, it is easier, as the other items in the study may be used to construct the match for the missing item.

c. Averages. Another technique frequently used is simply to substitute the average value for the respondents for the missing value. Note that this is essentially equivalent to the direct proration method.

d. Other techniques. Other ways of imputing that may be appropriate under given circumstances, for missing item data, would include the use of information from previous or similar surveys in place of the missing data, and the use of a similar (or proxy) variable in place of the missing one.

D. Weighting

It should by now be clear that one rarely uses the data collected from a survey sample as is. There are almost always a number of numerical adjustments that must be made. Collectively, these numerical adjustments tend to fall under a process called "weighting," in which the sample data are multiplied by a factor(s) designed to produce the best possible estimates of the population values for the study variables.

Generally, the various adjustments are combined into a single multiplicative factor called a "composite weight." The necessary adjustments are then accomplished by applying the composite weight to each unit of sample data to produce the final population estimates for the survey.

In some studies there may be several sets of weights required. For example, in the multistage sampling process, if one is to deal with data based on the first stage sample, that is, schools, the weights will differ from those that will be applied to the second- or third-stage data, for example, for teachers or students. This is because of the additional sampling, and different nonresponse characteristics associated with these stages.

In most cases at least the following adjustments are likely to be required as a part of the weighting process:

1. Sampling ratio

Adjustments will have to be made to prorate the sample data to the population. These may be complex, depending on the nature of the sampling process—multistage, differential sampling ratios, and the like. In any case, these adjustments will take the form of developing an effective or net sampling ratio that reflects all of the various levels of sampling and dividing the results for the appropriate segment of the sample by the corresponding sampling ratio that was used to select that sample segment.

2. Nonresponse

Further adjustments are required to impute for form nonresponse and item nonresponse (where appropriate). In addition, these computations may be complicated by the nature of the imputation chosen for implementation.

E. Some Further Comments on Errors

Although it remains important that survey reports include estimates of the errors characterizing the survey, the computation of the sampling errors is vastly complicated by the weighting procedures entailed in the various adjustments described above. I shall not try to go through this issue in detail, but consider the following.

Most sampling error estimates are a function of the number of cases involved (sample size). The application of the various weights to the sample data generally increases all of the apparent Ns, both for forms and for items. Obviously the data has not become more reliable simply because it has been weighted, but authorities often disagree as to how these changes should be incorporated into the procedures for estimating error.

A companion problem is that tests of statistical significance depend upon both error estimates and Ns. Thus, it is clear that for large surveys with complex designs, the basis for significance testing and confidence intervals tends to become tenuous.

We will pass the details of these computations and concerns, as I said, settling for making a point that the user of statistical surveys must be aware of these weaknesses in the overall process. We will also note that all of this notwithstanding, the user has the right to expect that the survey producer will provide *some* error estimates accompanying each survey. In addition, surveys should contain some guidance from the producer in order to assist the user in his interpretations of the data and error information presented.

One of the ways in which this problem is sometimes handled is to develop and present "generalized" variances and standard errors. These are error statistics that, although not precise with respect to a particular application, provide reasonable approximations for a range or group of similar statistics. They can be very useful in guiding interpretations.

V. A Tour Through Survey Country

In the following sections I am going to take you through a survey in more detail, as an illustration of the process that we have been discussing. Let us suppose that we have been commissioned to find out something about the schools and teachers in the *private* elementary/secondary education system in this country.

Most people are quite aware of the public schools, and indeed the U.S. Department of Education publishes much in the way of statistical information about the public school enterprise. Less is generally known and published about private schools, although they are believed to comprise about 25% of the nation's elementary/secondary schools, and educate about 12% of its children.

In these next few pages I shall frequently talk as though you are actually doing such a survey yourself, giving advice and tips about how to deal with certain things. Obviously many of you will never indulge yourselves in the joys and anguishes of survey work directly. But, equally obviously, the same points will apply to those surveys you turn to for information you want or need.

What I am suggesting is that these are points about which you may wish to satisfy yourself as you assess the quality and usefulness of the information you consume. I do not need to remind you that the shear fact that information has found its way into print most assuredly *does not* guarantee its accuracy or its value! A healthy degree of skepticism is always in order.

A. Problem Definition

The first thing to be considered is just what it is that we want to know—and from whom. It is not necessary at this stage to detail the questions to be asked, but it is necessary to know what topics are to be covered, as this may interact with the definition of the population, the sample, the nature of the instrumentation to be used, and other such factors.

We decide we want to know:

Information about schools:

- Religious orientation
- Enrollment by grade level
- Numbers of paid and volunteer staff
- Teacher remuneration
- Tuition
- Types of programs offered
- Years of operation
- Annual operating budget

Information about teachers:

- Education and training
- Years of teaching experience
- Full vs. part time
- Teaching assignment
- Hours spent on school-related activities
- Salary
- Opinions on educational goals and issues

How did we select these topics? Partly because they are of historical interest; partly because policy-makers frequently request such information in their decision-making processes; maybe because of our knowledge of some upcoming legislation or other event; maybe because of some specific mandate or request from our superiors or associates.

Obviously, these are not the only topics that could be surveyed; and, obviously, given that any survey has some limitation in terms of time to design and process, cost, and respondent tolerance, there will be competition for a place on our survey among these and other possible topics. In the design of the instrument(s) there will be more competition in terms of the proportion of the inquiry devoted to each selected topic.

For these reasons, it is always good practice to have a justification (preferably a written one) for each topic and query finally chosen for inclusion in the survey. The development of such a "rationale" not only justifies the inclusion of a topic on the survey, possibly at the expense of other worthy topics, but can assist in clarifying thinking about how such information may eventually be collected, analyzed, and used.

Some other things come clear from the exercise so far. It looks as though we might have to have two separate instruments, one for the schools (who should fill it out?), and one for the teachers. We must also decide what a "school" is in terms of a formal definition, and likewise, what a teacher is.

Let's consider the issue of what a school is. This is not as simple as it may sound. The crux of the problem has to do with whether or not to include some of the marginal possibilities. For example, do we include "home" schools—where, with the permission of the authorities, the parent conducts school for the family children (and sometimes for the neighbor's children) in a private home?

Should nursery schools be included—and can we distinguish them from daycare centers? How should we look at schools whose primary orientation is religious education? What about schools for "special and handicapped children?" A decision must be made with respect to each of these, and others. I should note that private schools may belong to any one of more than 20 different organizations—or no organizations at all—so that the matter cannot be addressed by simply saying we'll include all members of the National Catholic Educational Association or the National Association of Independent Schools, or some such.

Suppose we define an eligible school as one that meets the following three criteria:

- It has a first grade or higher.
- It is housed in other than a private home.
- It provides 4 or more hours/day of educational activities, for a minimum of 160 days a year.

Such a definition tends to eliminate daycare centers and nursery schools; it eliminates all home schools as well; it tends to focus on the standard range of elementary education. It does not address the schools that teach strictly a religious curriculum as opposed to those whose offerings are broader, however. This is because such a definition would require detailed program information—which is not generally available until after the survey is done. It does, however, eliminate the seasonal and the less-than-half-day programs.

(Whether reasonable or not, this definition has in fact been used by the U.S. Department of Education in one or more of its surveys of private education. The Department has, however, been criticized for avoiding the issue of pre-first grade education vs. daycare. Because this definition eliminates this entire area, Department of Education surveys of private school education have in the past resulted in "undercounts" by an unknown amount as compared to surveys carried out by the Bureau of the Census, which uses a somewhat different definition which includes some preschool education.)

I won't discuss this further, but suffice it to say that some of the most important thinking, research, and work is involved in the early stages of a survey in terms of defining just what the inquiry is all about, and to whom it will be addressed.

B. Sampling

Once we have decided what a school is, and what a teacher is, we must decide how to sample them for data collection. However, as pointed out earlier, in order to engage in the standard sorts of sampling procedures, it is necessary to have a list of the population from which to draw a sample. If we decide that we will draw our teacher data from teachers taken from the schools selected for the inquiry, we don't have to worry about an independent teacher list. This is a good thing, too, because at the time of this writing, there isn't any national list of private school teachers by any definition of teacher whatsoever!

1. The school list "frame"

A review of the possibilities shows that there is no accurate list of private schools available from public sources, but there are some private organizations that make a business of providing various lists for various business purposes. One of these is selected, and a list of some 26,000 schools is purchased.

Although the top organizations in the education listing business do a good job of updating and culling their lists, experience dictates that the true population of private schools will differ somewhat from the purchased list. This will be especially true if their criteria for defining a school differ significantly from yours, and also if it has been some time since their last update of their list. Thus, it is necessary to subject the purchased list to scrutiny.

The first step is to edit this list to remove entries that do not comport with the definition of school that we have chosen. Assuming that the necessary information for such a check is included in the list file, the file is run and schools failing to conform to the definition are purged.

Another step that could be taken is to compare this list against previous lists used for the same or similar purposes. Of course, if you have never done this survey before, or if the previous effort was several or more years old, or if lists obtained from outside sources do not conform to yours in important ways you are faced with having to make the best list you can on your own.

Now you have essentially established the "list frame," or population of schools for your survey. You may either try to collect data from them all, or draw a sample for data collection. If you choose to draw a sample, you will probably wish to stratify according to several more or less traditional stratification variables, such as school size, rural/urban location, and geographic location in the country (state or region).

After the sample is drawn and the data collection completed, you will get important information about your list frame. Among the schools that fail to respond to your inquiry you will find some that have closed or are disqualified in some way. Some of the ones that did respond may reveal reasons why they too no longer qualify. Such entries can be removed from your population list before conducting the analysis.

2. An "area frame"

The real problem, however, is undercoverage. That is, what about schools that exist, but were somehow overlooked in compiling the list frame? There is no convenient way to check for omissions. One possibility (which was used at one time by the U.S. Department of Education to supplement an edited and updated purchased list frame) is an "area frame."

In this instance you make the best list frame that you can, using whatever sources are most likely. Then, using census district data, you draw a sample of census districts for intensive study (probably stratified in a similar fashion to the list frame). The area sample is then all schools meeting the definition that can be found in the census districts comprising the area frame sample.

Because it is an independent sample which overlaps with the list frame sample, it becomes possible to compare the list frame for a particular area with the area frame for that same area, and thus to assess the completeness of the list frame. Based upon this comparison, statistical corrections can be made to the list frame, and survey statistics may be based upon joint data from the two frames.

To use the area frame in this way, it is assumed that each selected census district will be exhaustively combed for schools meeting the definition. Phone books and Yellow Pages will be consulted. Milk delivery routes will be checked. City and town directories will be examined, and local personnel of various types will be queried. Only by such intensive tactics can you be sure that you have identified all of the schools in the area that meet the definition.

(In passing, it should be noted that the use of two such frames for the survey study vastly complicates the process of weighting the data and constructing error statements for the results.)

3. The second stage sample of teachers

Once the school sample has been identified, we must draw the linked sample of teachers. If we have plenty of money, we can take all teachers in the selected schools. Otherwise, we will have to ask the principal to make a list of teachers and choose randomly among them, according to explicit directions supplied by us. (This is a weak spot, as there is no good way to know how well the principal followed instructions in selecting the teachers.)

The cumbersome alternative is to ask the principal to send us a list of teachers so that we may make the selection and notify the school as to who is to respond. In any case, we will probably include all teachers up to some number such as 5 and a random selection totalling some number like 5 from those schools with more than five eligible teachers. *It will be essential to know the total number of eligible teachers from which the selected sample was chosen.* Without this vital bit of information, it will be impossible to construct the appropriate weights for the sample data.

C. Designing the Data Collection

Although we settled on the topics to be covered by our survey (see above), decisions must now be made about precisely who the respondent is to be and how the data are to be collected. Then the appropriate data collection instruments (questionnaires, interview schedules, telephone protocols, or whatever) must be developed.

In our example, we will keep it simple and say that the school data is to be supplied by the principal and that the teacher data is to come from each teacher selected. We will recognize that although the teachers may actually supply their own data, the school data will in some cases come from the school secretary or an administrative assistant. But, we will ask the principal to sign off on the return, and thus will accept it as emanating from his/her level.

The choice of mail, telephone, computer-assisted-telephone-interview (sometimes called "CATI"), or personal interview as the data collection technique depends on time, money, and available skilled personnel. Mail is the cheapest and therefore usually offers the ability to cover more respondents. But, response rates are lower than with the other techniques, and, of course, the use of the printed format offers little or no flexibility in collecting the data.

On the other hand, personal interview is generally the slowest and most expensive, but offers high flexibility and the opportunity to judge intangibles that would be difficult to incorporate in the more rigid formats.

All of the real-time contact techniques are complicated to use if you are trying to collect information which must be looked up or assembled. If you must use telephone or interview for such situations, it is usually necessary to give advance notice of what is wanted so that the respondent may be prepared to provide it at the time of the contact.

There are other competing forces here. Flexibility is not an unmixed blessing. The ability to vary the order and specifics of the questions asked, in the interests of maximizing cooperation and rapport and following the logical flow of the interview, is very useful. However, such variations in the data collection routine tend to reduce the comparability of data from case to case and interviewer to interviewer, tend to reduce statistical reliability, and may produce misleading and invalid results. This can be offset somewhat by intensive training of interviewers, but great care must be taken.

In our survey, we have already determined that the area frame sample must be done on a personal basis because of its exploratory nature. If we are to take a national view of private schools, it is likely that the list frame survey will produce too many cases to be handled by other than mail questionnaires. In order to avoid too much detail here, I will discuss only the problem of developing the questionnaire.

1. Developing the instrument

a. The instrument plan. What we need first for our questionnaire development is an instrument plan. This plan will specify the type of questions, and their concentration with respect to the specific topics we identified earlier. In its fullest form, it will also include the specific rationales for the topics covered and the allocation of questions to those topics.

The Plan will also include question specifications. We must decide on fill-in-the-blank questions versus multiple choice questions versus free response questions. The former are best if reserved for numerical responses the respondent is likely to remember, like age, number of dependents, etc. Multiple-choice questions can be used for most things that we know enough about to specify a reasonable set of categories from which the respondent may choose. Free response (write-in) items must be used very sparingly because they are exceedingly time-consuming to score or process, and are often highly unreliable.

b. Tips on item writing

- If you can, ask your question as directly as possible. Don't depend on three other "related" questions that *might* provide the answer, but only with some more or less doubtful interpretation!
- Use "skip patterns" sparingly. Don't use "nested skip patterns" if at all possible. (Skips are instructions of the nature "if...then skip to Question_____." When you have skips within skips both the respondent and the analyst tend to get confused.)
- Avoid "double-barreled" questions (two or more questions wrapped up in one statement). You won't know which question the respondent really responded to!
- In choose-a-category, or multiple-choice, questions, make sure that the categories are not overlapping, so that a clear choice can be made.
- Make sure that each question contains a clear "problem" so that the respondent does not have to read your mind to answer. (This is particularly important in fill-in questions; e.g., "George Washington was _____?" A man? A surveyor? A general? A husband? The Father of our Country?)
- Remember to state for each question (or set of questions), where applicable, the extent to which estimates or approximations are acceptable.
- Questions should be as specific as possible. Specificity tends to enhance clarity and improve reliability.
- Although questions should be as succinct as possible consistent with clarity, never hesitate to use a few more words to remove ambiguity.
- Question formats and layouts should be logical, intelligible, and readable. Difficult formats lead to random responses or no response from the subjects.
- Question language should be kept to eighth grade level or less whenever possible.

Above all, remember that good question writing is a comparatively high level skill, not a "throwaway chore."

c. Some other cautions. The survey participant deserves to know the purpose of the survey and the uses to be made of the information collected. Such information should be included with every survey packet.

Directions and instructions must be of exceptional clarity. The capacity of the human organism to misunderstand approaches infinity!

Also, wherever practicable offer to provide the respondent some "quid pro quo" for his/her participation. This may take the form of a thank you letter, a certificate, a copy of the final report, a copy of the processed data for that specific responding entity, or some such. Be wary of actually paying money for responses, however. This can lead to all kinds of complications and biases. If responses are paid for, be sure to evaluate this factor, and also be sure that payment has full approval of all of the administrative authorities that may be involved.

2. Tryout

Once your questionnaire is in acceptable form it is ready for tryout. Tryout, or field testing, is an important step, particularly in surveys, where it is usually impossible to improve survey instrumentation after the fact of utilization. To tryout the instrument it is necessary to identify a small sample of the intended recipients of the instrument and administer it on a personal level.

This affords the opportunity of checking the goodness of the instructions, procedures, and expected timing of the instrument, as well as an opportunity to examine how well each question appears to function. If you did well in your design, it may be possible to go to the field (main survey) with modest revisions. If the revisions indicated were extensive, however, it would be well to run another tryout or field test on the revised version before going on to the main study.

You will read of reliability and/or validity studies conducted post-survey. I believe that such studies are superfluous and impractical wastes of time and money, if the design and field testing were properly conducted. In the design phase you should make sure that the data you are collecting is the best information on the topics to be covered that it is possible to get. In the field test you should assure that your survey design and procedures are adequate to obtain those data, with the highest possible reliability and validity.

If you have done those two things, there is nothing that an after-the-fact reliability study can do to improve the data you have already collected! You have what you have—and, it is highly unlikely that you will do the survey over under any circumstances, no matter what the post facto reliability analysis showed. Surely it is preferable to invest the time and money involved in beforehand efforts to improve the survey products than in after-the-fact hand-wringing about what can no longer be changed.

Of course the after-the-fact study might give some hints as to what to do different next time—but most of this should already be clear from your field test experiences, combined with your analysis of the actual survey process during the main study.

3. Data collection

a. **Secure informed consent.** Now we are ready to collect our data from the private schools. The first data collection step is to secure approvals and inform cognizant officials of the data collection in the offing. If you skip this step, you will run the risk of having whole segments of the respondent population denied to you by high level official decision.

For example, in the case of the private school survey, I would want to inform, and to the extent possible, secure the endorsement of the major private school associations and organizations. Many individual schools would seek guidance from these entities before deciding to participate, and others would refuse if they felt that your survey was contrary to official policy.

In some instances, depending on the nature of the survey and the respondent population, it is desirable or necessary also to secure the informed consent of the respondents themselves. This concern has been reflected in the various "privacy laws" which have appeared on the scene in recent years. This is a difficult area, and any survey sponsor must be fully informed of the latest requirements for approvals, consents, the maintenance of personal identifiers and records, confidentiality, and similar privacy-related concerns.

b. **Document control.** Having mailed our survey at last, we must deal with the flow of the survey documents. There are three major activities required in the data collection process:

(1) Monitoring and tracking. It is absolutely essential to know exactly to whom you have mailed what. When returns start to come in, you must know from whom you have received what. These days, mailing labels are often machine produced and affixed, and computers employed to keep up-to-the-minute records of the mailing process. Receipts must be logged in and similarly monitored.

(2) Follow-up. Not every school will respond to the first mailing. At some point in time, 2–4 weeks usually, after initial mailout, some follow-up of nonrespondents is in order. (Here it is essential that your monitoring and tracking is good enough to enable you to know that a nonrespondent has really not responded.) Follow-up may take the form of a reminder of some kind, or the remailing of the questionnaire. Some research suggests

that remailing is more effective than a postcard, but personal contact is probably more effective yet.

You continue to track returns, and, depending on flow of returns, money, and time, you may go through several follow-up cycles. At some point, however, you must declare the data collection cut off and concluded.

(3) Editing and data entry. These activities are usually performed by computer these days, though there is much to be said for a personal scan of at least a selected subset of the data forms. Such a scan can pick up pattern marking (marking according to a pattern, such as zigzag, instead of the intent of the questions), oddities, and distortions that would be difficult to pick up by machine.

Computers can perform checks to test the responses for consistency, out-of-range answers, presence/absence of key items, and the like. Sometimes fail-edit forms are returned to the follow-up cycle in some way, sometimes they are subjected to special telephone follow-up, sometimes they are rejected entirely, and sometimes they are flagged for special treatment during processing.

Edited forms are then usually entered into machine-processable files. This step is rarely done any longer by keypunching IBM cards. It is usually done by key-to-tape or key-to-disk machines, sometimes with further edit procedures built into the process. Some forms, if they have been designed in advance for it, may be processed by optical scan machines so that the information on the form is read directly into machine-processable files. In general mail questionnaires are entered directly into the data base after passing edit procedures.

Although we have been discussing the mail survey in particular, I should digress to mention that there are some additional steps that may be necessary if data collection procedures involving interviewers are used. I have mentioned earlier the importance of training the interviewers to be as knowledgeable as possible, and as similar to each other as possible. The similarity is necessary so that their data may be pooled in the study without being subject to specific interviewer effects (biases) that would distort the respondents' information. Before entering the interviews into the data base, it may be desirable to review some of them to assure that the interviewers have been producing consistent and comparable results.

Ideally, after training, interviewers should be field tested, just as you would field test an instrument, and retrained as necessary. The forms used by interviewers to record their data are similar to questionnaires and should be a part of the field test. In a sense, we are talking about an "instrument" that is really a composite, comprised of the interviewer and his data collection form functioning jointly as a system. It is the reliability and validity of this system with which we are ultimately concerned. Data derived from this system is then subjected to tracking and monitoring, follow-up and editing and data entry, just as it would be for questionnaire data.

4. Data analysis

Once the editing process is finished, and the resulting data has been entered into machine-readable files, you are ready to start the process of manipulating the data to yield information. If you are a good researcher, you will have created an "analysis plan" during the early stages of your survey planning. Unless you have done so, you might now find that certain key elements of information cannot be derived from the data you have collected.

This happens quite frequently. Sometimes it happens because the original reasons for the survey were illogical, unknowledgeable, ambiguous, or the like. Sometimes the design work and field test fail to reveal misconceptions or fallacies and key elements of information don't exist or are not supplied by the respondents. Sometimes people just cut corners, for reasons of time and/or money or carelessness.

At any rate, it is the analysis plan that should have guided the specification of the actual data elements to be collected. Now that these data elements are on hand, the data analysis is a simple matter of following the plan.

Often the plan calls for some "pre-processing" of the data. This may be done in order to modify the data in particular ways to make the main processing easier or more accurate. For example, if we want age of teacher, we may have collected it in terms of years and months because we felt that respondents would find it more natural that way. Now, for analytic or processing purposes, we may wish to create an "age" variable in terms of years and fractions of years, or perhaps total number of months.

Another example would be the creation of scales from individual items; if we had several opinion questions about the teachers' feelings toward their employment situations, we might wish to create a composite of these items on the subject of job satisfaction, or adequacy of facilities available, or some such. Such activities would be a part of the preprocessing.

Another aspect of the preprocessing would be the creation of a file of "weights" to be applied to the raw data to create national projections and estimates.

The main data analysis would consist of the computation of sets of statistical results for the various items and scales, weighted to reflect national projections and estimates. That is, we don't much care how many teachers were in the sample, but we would like an estimate, with a confidence level, of how many there are in the nation.

The analysis would provide weighted frequencies and percentages for categorical items such as "primary teaching assignment" and means, standard deviations, Ns, standard errors of the means, ranges, etc., for the continuous variables such as age, years of teaching experience, and the like.

Generally, each item of collected data would be statistically analyzed, and the results presented along with some kind of error estimates (confidence intervals, standard errors, generalized variances, or coefficients of variation). Such results are usually presented as books of tables, complete with labels and legends, so that the user can be aware of just what he is looking at. If not too bulky, they often comprise the Appendix of the Report which describes the study procedures, and often offers insights on the meaning of the findings.

D. The Survey Report

There are some things that are nice to have in a report—and some that should be regarded as necessary. It is nice to have an "executive summary" in the beginning of the report. This is a piece, usually quite brief, that offers a summary description of the essentials of the survey. It should include a synopsis of the methodology and a summary of the most important findings. It is also nice to have a table of contents and an index, so that the user can find things he is interested in—which quite frequently is not the whole report in its entirety.

It's also good if the report is readable—adequate grammar, not too verbose, well-organized, and to the point.

The things that I regard as essential in a survey report are the things which make it possible for the user to decide for himself how applicable and dependable the information presented is for his/her own purposes. This requires the following:

1. Purposes and procedures

A clear statement of the purposes of the survey, and the methodology and technical characteristics of the survey employed to achieve these purposes. This must include the following items:

- A complete description of the population to which the survey data may be legitimately generalized. For example, are the data for years of teaching experience applicable to all teachers, only male teachers, only those in the Northeast, etc.?

- A description of the sampling methodology, including sampling ratios, stratification, and response analyses.
- A general description of the data collection procedures, particularly any special or unusual aspects, and how well they worked.
- A description of the weighting process.
- A general description of the editing, data handling, and data processing procedures.
- A summary description of any limitations imposed by technical factors on the survey data and their interpretation and the expected impact of such limitations in the opinion of the study authors.
- A reference list of technical authorities for the procedures used if these are arcane or controversial in any way.

(Depending on the complexity of the study, the technical material may be too bulky and involved to be included in the report proper. If this is the case, the material may be summarized in the report, and the complete discussion relegated to an appendix.)

2. Findings

A presentation of the data developed by the survey is required. Frequently the complete set of data tables produced by the data analysis activities is too extensive for inclusion in the report proper, and must be presented in an appendix, or even in a companion volume. In these cases, it is appropriate to include summary tables pertinent to the more important findings as part of the text, referring the reader to the appropriate supplementary source for the complete set of findings.

(Note that it is not necessary—nor desirable—to include in the report every scrap of data analysis that might have been conducted. If you have followed a data analysis plan, focussed on substantive issues, you will have avoided the temptation to "run everything by everything else." I hope so. Even so, the data tables chosen for inclusion in the report should exhibit some rationale related to the purpose of the survey, and should not necessarily be exhaustive of all possible analyses of the data.)

3. Conclusions and interpretations

Some comments or conclusions are in order, either highlighting important and/or unexpected findings in the data, or drawing the users' attention to the more important facets of the data. There is an implication here that you should be prepared to pass these data through the organic computer between your ears. After all, in some ways you are the world's greatest living expert—at least on this particular body of survey data! **What is**

absolutely essential, however, is that it be perfectly plain to all users just what is your *opinion* as opposed to what are your *data*. Also, try not to go too far beyond your data in offering your interpretive analysis!

4. Summary

A final summary is very useful. This should draw together the most important findings and conclusions, state implications, and discuss what the impacts have been of any recognizable limitations of the survey and its methodology. It is permissible, though not mandatory, to state recommendations for further work, and for improved survey procedures, in the area of interest at this point.

5. References

A list of such references and authorities as may be required to support any interpretive comments and conclusions that go beyond the obvious level of the data is highly desirable.

E. At Tour's End

If you have taken seriously your professional obligations to develop and conduct the best, state-of-the-art survey possible, we will have gathered a great deal of data about private school education which may be of high value. If you have presented this information in a report that conscientiously attempts to present the information in an unbiased, scientific way, but one that permits the informed user to interpret and apply the information for his own purposes, you will have made a significant contribution to American education!

VI. Summary

A lot of ground has been covered in this chapter. However, the focus has been on sample surveys. Surveys tend to be large-scale data collections about people. For many reasons they tend to be based on samples of the population of interest rather than censuses (which include everyone). The vast majority of surveys are of the sampling type.

I noted that there are several kinds of samples, but that the key characteristic of any sample, of any kind, is that it must be as representative as possible of the population about which information is sought. Most sampling procedures depend on random process to attain representativeness, but since random process does not guarantee representativeness for any individual sample, it is necessary to develop statistical estimates of sampling error to estimate the goodness of population estimates based upon samples.

After discussing sampling errors, I then turned to the other major category of errors encountered in the survey process—nonsampling errors. These are errors that arise because of systematic biases originating in the survey methodology, its instrumentation, its processing, or in the population or sample definition. Although statistical estimates of sampling errors can be made in most cases, nonsampling errors are more difficult to detect and correct. We run an especially important risk of nonsampling errors when we "impute" for missing data (nonresponse in items or forms).

As has been mentioned before, we are rarely interested in sample survey data for themselves, but rather in population estimates based upon them (generalizations). To transform sample statistics into population estimates it is necessary to multiply them by "weights" that reflect the appropriate sampling ratios and imputations for each sample statistic. This greatly complicates the computation of statistical estimates of sampling error. Nonetheless, some estimates should always be reported wherever possible. These might include confidence intervals, standard errors, generalized variances, and coefficients of variation.

Finally, we closed this chapter with a "tour of survey country" in which I recounted briefly the various steps and procedures entailed in doing a survey of private education. Cautions and hints about good procedure were included for your use, or, more likely, as a basis for your critical appraisal of survey data with which you come in contact.

VII. References

Cocheran, W. G. *Planning and Analysis of Observational Studies*. New York: Wiley, 1983.

Henry, George T. *Practical Sampling*. Newbury Park: Sage Publications, 1990.

Kalton, Graham. *Introduction to Survey Sampling*. Beverley Hills, CA: Sage, 1983.

Lavrakas, Paul J. *Telephone Survey Methods, Sampling, Selection, and Supervision*. Beverly Hills: Sage Publications, 1987.

Manly, Bryan F. J. *Stage-Structured Populations: Sampling, Analysis, and Simulation*, New York: Chapman & Hall, 1990.

Rea, Louis M. and Parker, Richard A. *Designing and Conducting Survey Research. A Comprehensive Guide*. San Francisco: Jossey-Bass, 1992.

Schaeffer, R. L., Mendenhall, William and Ott, Lyman. *Elementary Survey Sampling, 4th Ed.* Boston: PWS-Kent, 1990.

Singer, Eleanor and Presser, Stanley (Eds.) *Research Methods: A Reader*. Chicago: University of Chicago Press, 1989.

10

General Summary and Epilogue

It would never do to attempt to summarize everything covered in the preceding nine chapters. However, perhaps some of the overarching ideas bear recapitulation. And, I think I should restate my goals as I now see them. Let's take the latter first.

The first and foremost purpose of this book is to impart and extend understanding of things statistical. In part I have tried to reach this goal by peeling away some of the jargon, the complex mathematics, and the concern with derivations and formulae that befog the typical statistics book.

I don't expect that each reader will fully understand all of the topics covered—indeed, there has been neither time nor space to treat any topic in sufficient detail and depth to ensure uniform understanding. But, I would hope that every reader comes away from his reading with at least some understanding that he or she did not possess before.

To achieve this major purpose, I have tried to place statistics on a foundation of conceptual and numerical theory and experience. That required showing the reader something of the place of statistics in the scientific world, the kinds and sources of data, the nature of experimentation, the nature of numbers, and so forth. In addition, because statistics is built on numbers, I thought it necessary to review the basics of number usage and manipulation, at least in those areas drawn upon by statistical procedures.

My secondary purpose was to emphasize the idea that statistical ideas and concepts, far from being rare and academic, pervade everyday life for most of us, in one form or another. Not only do we all estimate the consequences of our behavior and predict the behavior of others, but we all encounter statistical information, polls and surveys, and the impact of their results, on a daily basis. I would venture that a majority of us must deal with (consume) statistical data of some sort in our occupational

334

settings as well. Thus, any improvement in statistical understandings—computation aside—can be of potential value to most of us.

I. Some of the Major Notions Presented

Following are some of the major notions presented in the body of the book:

- The primary definition of statistics is that it is a body of principles and procedures intended to assist in eliciting useful information from masses of data. Statistical procedures rarely apply to individual cases.
- There are three principal uses for statistics: to describe the information inherent in a body of data; to infer unknown population values from data based on parts (samples) of that population; and to describe the relationships among two or more variables, based on data about each. Statistics is applicable whenever an enterprise generates large quantities of (usually numerical) data, such as in the sciences, social sciences, business, education, and quality control.
- The development of statistical procedures has depended heavily on several fundamental ideas, which are familiar to most of us in everyday life (although their transformation into statistical computing routines may be very obscure). These ideas include: averages, approximations, variability, interpolations (proportioning the difference), randomness (chance), and probability (the likelihood that something will happen).
- As is true of science in general, sound statistical results depend in the first instance on sound conceptualization of the problem under study, including the study design, control of error, and the development of data with both reliability and validity. Such data may then be processed using appropriate statistical procedures to assist in extracting and interpreting the information therein.
- In describing the information contained in a body of data, there are three principal statistical techniques that are commonly applied. The first is to study the distribution of the data. This is done by forming an array of the data, frequently in the form of a frequency distribution or grouped frequency distribution. Such arrays can be plotted, or graphed, so that they may be studied pictorially, and possible compared to one or more of certain standard "curves" in terms of modality, symmetry, normality, skewness, and kurtosis.
- The second step is to examine the distribution for central tendency, and to compute an index of the location of central tendency, assuming that central tendency is extant. Such indices are termed averages. The two most common and useful are the arithmetic mean and the median. The

first is defined as the sum of the values divided by the number of values; the second is defined as the score point below that exactly 50% of the values in the array may be found. Such indices may be calculated on any distribution, whether or not grouped, and whether or not plotted.

- The third descriptive step is to measure the extent to which the distribution is spread out as opposed to being bunched up—its variability. The most common measure is the standard deviation, a companion to the arithmetic mean. It is defined as the square root of the arithmetic mean of the squared deviations of the values from their own arithmetic mean —or, the quadratic mean deviation from the distribution's arithmetic mean. If the square root is not taken, then the quantity is called the "variance." The companion measure to the median is usually the semi-interquartile range, defined as the score range within which the middle 50% of the distribution may be found, divided by two—a sort of average quartile.

- If data for a set of items, people, or events is collected on *two* measures, or variables, it becomes possible to ask how the two variables may be related. The data set can be plotted jointly on a graph with one variable along the *x*-axis, and one along the *y*-axis, and points on the graph representing the simultaneous *x* and *y* measurements for each of the items in the data set. If such a "scatterplot" is based on a fairly large data set and the joint distributions are in grouped form, it becomes possible to devise a way of "regressing" one variable on the other, that is, of predicting one variable from the other. In general, the best *y* prediction is given by a straight line of best fit drawn through the *y* means of each of the *x* categories—and vice versa. This is called "linear regression."

- The usual index of the strength of relationship, or the goodness of prediction, is the "correlation coefficient, '*r*'." The value of this index ranges from 1.0, where the two variables are perfectly related; through 0.0, where they are totally independent of each other; to -1.0, where they are again perfectly related, but in an inverse way. The value of the index may be defined variously, but one of the more useful ways to think of it is in terms of the errors made when the predicted values are compared to the actual values of the predicted variable. The index is the square root of the difference of the total variance in the predicted variable less the error variance in the prediction.

- Strictly speaking, Pearson product-moment correlation coefficients (as described just above) apply only to two variables that are linearly related. However, approximations have been developed that apply to ranks, categorical data on one or both dimensions, curvilinear situations, and prediction based upon multiple predictors.

- There is a great tendency to confuse correlated variables with causal situations. In reality there are four causal possibilities when correlation between two variables, A and B, has been demonstrated: A caused B; B caused A; A and B were both caused by an unknown C; and the observed relationship between A and B was just a fluke (chance). It is not possible to know with which of these four cases one is dealing without additional information beyond the correlation coefficient.
- The question of statistical inference arises when we attempt to estimate the value of some unknown population measure from data based on a part, or sample, of that population. We may make the attempt as a descriptive effort, in which case it is usually called a "point estimate." Or, we may have some hypothesis we wish to test with the empirical data, in which case the problem is one of comparing the sample result to the proposed hypothetical value. Such hypotheses may be quite complex, involving the differences among various values, etc. However, in *no case* can statistical procedures *prove* a hypothesis. This is because the definitive proof may lie in the unknown, unmeasured portion of the population.
- Because hypotheses cannot be proved, statistical hypothesis testing consists of setting up a contradictory hypothesis, called the null, and attempting to discredit the null—thus, indirectly supporting (not proving) the hypothesis of interest.
- There are two ways to go wrong in statistical hypothesis testing: You can reject the null when it is true (alpha, or Type I error), and you can accept the null when it is false (beta, or Type II error). Logically, both types of error are important, depending on the problem, but traditionally, only the alpha error, also called the "level of significance," is controlled in most statistical problems.
- Control consists of transforming the hypothesis comparison into a value that has a known probability distribution and assessing the likelihood of observing the obtained sample result from a population in which the null was really true. If this likelihood is small (usually 5% or 1%) we run explicitly that risk of false rejection of the null and declare the test "significant." This means that we accept the hypothesis of interest with the indicated probability of error.
- The nature of the tests that can be made is limited by our ability to devise transformations of the measures into variables with known probability distributions, so that we can state the probability of alpha error. There are several such distributions in common use, including: the binomial, the normal, the "t," the "F," and the chi-square.

If more complex tests of distributional means are required, we turn to a set of procedures known as analysis of variance (ANOVA). These are based on the idea that the total variance in a set of measures can be partitioned into additive components, each attributable to a feature of the study design. Then the variance around the various submeans in the problem can be compared to other independent estimates of the population variance with the result that probability statements about the differences in the means can be made. If necessary, undesirable differences in the various subsamples in the problem can be statistically eliminated from the comparisons by use of a companion set of techniques called analysis of covariance.

- Error considerations also apply to point estimates of population statistics. Using very similar reasoning to that employed in hypothesis testing, probability distributions are used to set up a range of values with a known probability of including the unknown population value. This range is based on the sample data and probabilities derived from one of the standard distributions. In effect, the confidence interval states that the sample-based interval will bracket the unknown population parameter with a stated confidence (probability).
- Sometimes estimates are reported along with "standard errors." In this case, the confidence interval is based on the normal curve, and states that the estimate plus or minus 1 standard error will bracket the true value with a probability of .68 (the standard deviation points on the normal curve are approximately 34% on either side of the mean).
- Surveys are a very common part of today's life. Some are censal— collecting data from each member of a population. The vast majority, however, are of the sampling variety, where data is collected only from a small fraction of the population of interest. There are many reasons for sampling, including time, money, and accessibility.
- Sample data are of little interest in and of themselves. Their value is as a basis for generalizing findings and estimates to the population of interest. To permit generalization, samples must be as representative as possible. This means they must be as like the population as possible in all essential respects. We can control a few of these aspects in the sampling process, but to a large degree we depend on the operation of random chance to allot most of these factors to our samples in the same general proportions that characterize the population.
- Random process, however, does not guarantee that any particular sample will be representative. It only ensures that innumerable repetitions of the process would converge on the population characteristics. Thus, any particular measure would display a distribution of values if taken over repeated samplings. Some such distributions, called sampling distributions, are used to estimate the sampling error associated with the obtained sample value of the measure.

- Sample results can also be in error as a function of "nonsampling" errors. These errors arise whenever there is a systematic, nonchance bias that affects the results. They are hard to control, particularly because many may be unknown. One of the most common affecting surveys is nonresponse bias. When some percentage of the sample fails to return the form or answer a question, it is dangerous to assume that they are just like those who did and simply prorate (inflate) the sample results. The nonrespondents are most likely systematically different from respondents in important ways.
- When simple inflation of the sample results to population estimates seems unwise, it is sometimes possible to impute for the missing data. This may be done on the basis of matching the missing cases with similar respondents, assuming that the data for the match is available. Sometimes a special nonrespondent study is done. Proxy variables may be used. Sometimes, if the nonresponse is more than 20–25% and cannot be suitably estimated, it may be necessary to consider junking the study.

II. Conclusion

By now you may have concluded that there is more to those statistics that increasingly are found in reports, magazines, and the newspapers than you thought. But, by now, I hope that you have also acquired a richer understanding of the processes and procedures that give rise to such information, making possible some skepticism of the often extravagant and misleading claims that frequently accompany such statistics. A little skepticism, particularly if based on some understanding, is healthy.

Index

JOHNNY BLAIR

JOHNNY BLAIR